About Island Press

Since 1984, the nonprofit organization Island Press has been stimulating, shaping, and communicating ideas that are essential for solving environmental problems worldwide. With more than 1,000 titles in print and some 30 new releases each year, we are the nation's leading publisher on environmental issues. We identify innovative thinkers and emerging trends in the environmental field. We work with world-renowned experts and authors to develop cross-disciplinary solutions to environmental challenges.

Island Press designs and executes educational campaigns, in conjunction with our authors, to communicate their critical messages in print, in person, and online using the latest technologies, innovative programs, and the media. Our goal is to reach targeted audiences—scientists, policy makers, environmental advocates, urban planners, the media, and concerned citizens—with information that can be used to create the framework for long-term ecological health and human well-being.

Island Press gratefully acknowledges major support from The Bobolink Foundation, Caldera Foundation, The Curtis and Edith Munson Foundation, The Forrest C. and Frances H. Lattner Foundation, The JPB Foundation, The Kresge Foundation, The Summit Charitable Foundation, Inc., and many other generous organizations and individuals.

The opinions expressed in this book are those of the author(s) and do not necessarily reflect the views of our supporters.

A New War on Cancer

A New War on Cancer

THE UNLIKELY HEROES
REVOLUTIONIZING PREVENTION

Kristina Marusic

 ISLANDPRESS | Washington | Covelo

Library of Congress Control Number: 2022946105

All Island Press books are printed on environmentally responsible materials.

Manufactured in the United States of America
10 9 8 7 6 5 4 3 2 1

Keywords: BPA (bisphenol A), breast cancer, carcinogens, chemical exposure, childhood cancer, community health, environmental justice, green building, pesticide exposure, pollution, public health, race for the cure, The Toxic Substances Control Act (TSCA), toxic chemicals, women in science and medicine

For my family, both given and chosen

Contents

Foreword

Many types of cancer are on the rise in the United States. From 1975 to 2019, the number of new cancer cases per 100,000 Americans—the incidence rate—increased for multiple cancers. Incidence of multiple myeloma rose by 46%, incidence of non-Hodgkin's lymphoma by 76%, and incidence of testicular cancer by 70%. In the same years, incidence of childhood leukemia increased by 35% and incidence of childhood brain cancer by 33%. These increases are far too rapid to be of genetic origin. They cannot be explained by better diagnosis.

In the same years, cancer death rates dropped and survival improved—in some cases dramatically. These gains are the result of screening, early detection, and better treatments. They are major victories in the war on cancer. But the rise in cancer incidence threatens to undo those gains.

The explanation for the increasing incidence of cancer lies in our world of chemicals. Since the dawn of the chemical era in the early twentieth century, more than 300,000 new chemicals have been invented. These are novel materials that never before existed on Earth. Many are made from oil and natural gas. They are manufactured in enormous quantities, and global production is on track to double by 2030.

Some manufactured chemicals have greatly benefited human health. Disinfectants have brought safe drinking water to millions and reduced deaths from dysentery. Antibiotics prevent deaths from once fatal infections. New chemotherapies cure cancers.

But manufactured chemicals have also caused great harm. They pollute every corner of the planet from the deepest ocean trenches to the high Arctic. They kill bees, birds, fish, and mammals. The chlorofluorocarbon chemicals used widely as refrigerants came close to destroying the stratospheric ozone layer that protects all life on Earth against solar radiation. Manufactured chemicals enter people's bodies through our air, our water, and our food, and several hundred of them can be found today in the bodies of almost all persons on Earth, including infants and children. Some will persist for centuries. Chemical pollution has become so widespread and complex that in 2022, an expert body at the Stockholm Environmental Institute concluded that chemical pollution now exceeds our ability to monitor and contain it and thus threatens the sustainability of human societies.

Many manufactured chemicals cause cancer. Benzene, 1,3-butadiene, and ethylene oxide cause lymphoma and leukemia. Formaldehyde causes lymphoma and respiratory cancers. Vinyl chloride causes cancer of the liver. Benzidine causes bladder cancer. Exposure to the pesticide DDT in infancy is associated with increased risk for breast cancer in women in middle age. The World Health Organization has determined that more than 100 manufactured chemicals can cause cancer in humans.

The root causes of this chemical crisis are the repeated failure of chemical manufacturers to take responsibility for the materials they produce and the systematic failure of governments, including our own, to regulate toxic chemicals.

Scores of new chemicals are brought to market every year with great enthusiasm but with almost no assessment as to their possible dangers. Unlike prescription drugs and vaccines, which are carefully screened for safety, most widely used chemicals have never been tested for safety or toxicity, and fewer than 20% have ever been examined for their potential to harm fetuses, infants, and children. Most of the manufactured chemicals that are known to be human carcinogens are still sold today, and only five hazardous chemicals have been removed from US markets in the past fifty years. Chemical policy in this country is broken.

This powerful book by Kristina Marusic tells in stark yet very human terms how chemical pollution has silently infiltrated our lives and become a major threat to our health and the health of our children. But rather than dwelling on the enormity of the problem, this book details the lives and work of people who are advancing solutions, giving readers reasons to stay hopeful and new ways to push for progress.

Drawing on her skills as a reporter, Ms. Marusic renders thoughtful portraits of heroes across America who have devoted their lives to preventing cancers caused by chemicals: An Indian American researcher who grew up in the rural South and now fights racial injustice in toxic cosmetics through her work in New York; a Nigerian American children's health advocate making daycares and playgrounds across the nation safer through her work in Washington, D.C.; a California-based lawyer-turned-rabble-rouser driving the proliferation of carcinogen-free buildings through a nonprofit; and an activist living on the fence line of a Pennsylvania steel plant fighting to defend her community from toxic pollution, among others. Their stories are interwoven with that of a young woman who developed cancer after having lived most of her life surrounded by oil fields and petrochemical plants, highlighting the motivation behind all this work—protecting people from the hardships that accompany diagnosis of this deadly but preventable disease.

As Ms. Marusic says, it is time to launch a new war on cancer. The goal of this new war must be to not only treat and cure cancer, as we have done until now, but to prevent cancer by preventing exposures to the toxic chemicals that are its causes. We know how to do this, and we have done it before. It is time to act.

—Philip J. Landrigan, MD, MSc, FAAP

Introduction

Madelina DeLuca was diagnosed with leukemia about a month before her second birthday.

"She had some bruising and we couldn't figure out where it came from," her mom, Kristin DeLuca, told me in 2019 when I interviewed her for a story about environmental exposures and cancer.

Madelina's doctors ran a bevy of tests, but they didn't reveal any answers. Then Madelina developed inexplicable stomach pain, and the dark blue and purple blooms on her skin multiplied. After several months of uncertainty, a blood test came back showing too many white blood cells. In November 2014, her doctor ordered a bone marrow biopsy. "She got a fever after the biopsy, so they kept her in the hospital as a precaution," Kristin said. "Then it all happened really fast. Her test results came back the next day, she was diagnosed with acute myeloid leukemia, and within two days she had a port in her chest and had started chemo. It was very overwhelming."

In 2015, while Madelina was undergoing treatment at the Children's Hospital of Pittsburgh, a photo of her embracing another young cancer patient went viral. The image of the little girls—one dark-skinned, one pale, both wearing jammies, their nearly bald heads touching while they held each other and gazed out a hospital window at the city skyline— was shared around the world and garnered national media attention.

The other girl in the photo was five-year-old Maliya Jones. Her mother, Tazz Jones, captured the candid shot and captioned it, "This is the perfect example of love," in a Facebook post. In media interviews at the time, Tazz expressed hope that the girls would stay friends as they got older and would have the photo to commemorate the difficulties they had both overcome.

Maliya died a year later at the age of six.

Kristin told me, "Throughout the course of Madelina's treatment, she made a lot of friends who weren't able to defeat it."

Climbing Cancer Rates

The United States is more than fifty years into its "war on cancer," but the disease is still prevalent. Half of all American men and one in three women can expect to get some type of cancer diagnosis in their life-times. We're better than we've ever been at curing and treating the dis-ease, but cancer still claims the lives of one in every five American men and one in every six women. Rates of some types of cancer have fallen over time, but others continue to increase.

Childhood cancer rates are particularly striking. While cancer is still relatively rare among childhood diseases, Maliya and Madelina are far from alone. One in every 285 American children receives a cancer diag-nosis before the age of twenty, and cancer rates for children and teens across the US have increased steadily over the last fifty years. Rates of childhood leukemia have increased by 35%, and rates of childhood brain cancer have gone up by 33% since researchers started tracking the disease in the early 1970s. Cancer rates in children and teens across the globe have followed a similar upward trajectory.

This increase is too rapid to be the result of genetic changes alone, which would happen over centuries, not decades. Rapid increases in disease rates are sometimes explained by improvements in diagnostic capabilities. When our ability to detect a disease improves, we start finding more of it. This might account for some of the increase in childhood cancers, but it doesn't fully explain what's happening here. While many new tests enable us to learn more about cancer subtypes and make more accurate diagnoses, the fundamental diagnostic tools for the most common types of childhood cancers haven't changed. So if increasing childhood cancer rates aren't explained by genetic changes or better diagnostic tools, what's been causing such a rapid increase for the last fifty years?

Dr. Margaret Kripke, a leading expert in the immunology of skin cancers and professor emerita of the University of Texas MD Anderson Cancer Center, served multiple terms on the US President's Cancer Panel—a three-person panel that advises the president of the United States on high-priority issues related to cancer and oversees the development and implementation of the National Cancer Program.

In 2008, when she learned that the President's Cancer Panel would investigate environmental causes of cancer, Kripke bristled at the idea. A single study from the 1980s indicated that pollution and chemical exposures caused just 6% of all cancers, and that statistic had been widely accepted ever since. Kripke had never questioned the findings, so she thought the group's time would be better spent researching other aspects of cancer. But what they found during their two-year investigation left her stunned.

The panel determined that up to two-thirds of all cancer cases are linked to preventable environmental exposures (a category that includes any factor originating outside the body and our own DNA like smoking, pollution, and chemical exposures). They also learned that while around 80,000 chemicals are used in products sold to American consumers, fewer than 1% have ever been tested for toxicity or safety, and existing regulations on cancer-causing chemicals in consumer products are rarely enforced. "I had a set of assumptions that most people have, that chemicals are tested before they're put on the market, that things known to be carcinogenic are regulated, and that if regulations exist, they're enforced," Kripke said. "It turned out none of that is true. It was a completely eye-opening experience for me."

In 2010, the Panel released its report on environmental exposures as causes of cancer, marking the first time the subject had been covered in the organization's forty-year history.[1] Kripke explained that while individual cases of cancer cannot be traced to harmful exposures, a substantial body of research has shown that among groups of people, less exposure equals fewer instances of cancer. Fewer cancer cases occur in communities where there is reduced exposure to cancer-causing chemicals.

"Many of the researchers who came to testify before the President's Cancer Panel were incredibly frustrated because they had been saying this for a very long time, but nobody had been listening," Kripke said. "That was very moving to me, and it struck me as something that was quite amiss with the field."

The report was controversial at the time. Because of the difficulty of linking any one person's cancer to chemicals, some of Kripke's colleagues wanted to continue to focus exclusively on changing individual behaviors, such as diet and exercise, which she referred to as a "blame-the-victim approach to cancer prevention."

"I heard from a young man who was just completely outraged that he'd gotten cancer despite having a perfectly healthy lifestyle," she said. "He exercised, he never smoked, never drank, but he still developed bladder cancer. He felt like he'd been lied to. Having a healthy lifestyle is important, but it's misleading to let people think they can control whether or not they get cancer just by having healthy lifestyles."

Awareness about environmental exposures and cancer risk has increased since the report came out in 2010, but the fundamental problem persists. "There's still a lot to be concerned about in the way we do business in the US, where we use a reactionary principle rather than a precautionary one—meaning we wait until there's evidence that chemicals are causing harm before trying to regulate them," Kripke said.

In other parts of the world, including the EU, Denmark, and Sweden, regulators evaluate potentially harmful substances before approving their use, at least in theory. In the US, Kripke said, class action lawsuits seeking compensation for people who've been sickened or killed by chemicals often precede regulations, or even stand in for them.

Lawsuits are often reserved for extreme cases of exposure, but in fact, we're all regularly exposed to low doses of cancer-causing compounds in food and water, personal care products, and air pollution. We consume chemicals in the things we eat and drink, absorb them through our skin, and inhale them in the air we breathe.

Kids and teens are much more vulnerable to these constant, low-dose exposures than adults, according to Dr. Phil Landrigan, a pediatrician, public health physician, and epidemiologist who serves as director of the Global Observatory on Pollution and Health at Boston College. Kids breathe more air, eat more food, and drink more water per pound of body weight than adults, so carcinogens in their air, food, and water wind up in their bodies at higher concentrations. Because kids' defense mechanisms aren't yet mature, Landrigan explained, their bodies also have less ability than adults' bodies to remove things that shouldn't be there. And they're still undergoing complex development processes in their brains, immune systems, and reproductive systems that make those bodily systems more vulnerable. Thousands of steps must occur in precise sequence for healthy human development; if something gets into a child's body that disrupts those processes, even at a very low dose, it can initiate the development of disease, making kids the proverbial canaries in the coal mine.

Alarmingly, research also indicates that parents' exposures to cancer-causing chemicals—even before a child is conceived—may increase children's cancer risk.[2]

After accidents, cancer is the second leading cause of death in American children ages one to fourteen today. At the beginning of 2022, the American Cancer Society estimated that about 10,470 American children under the age of fifteen would be diagnosed with cancer the following year, and that 1,050 children would die from the disease.[3]

Meanwhile, rates of several types of cancer strongly linked to chemical exposures have also been on the rise in adults, as has the share of lung cancer cases showing up in people who've never smoked. And during the same time period, people of increasingly younger ages have started getting cancers typically seen in adults.

An Ounce of Prevention

From 1988 to 1992, American scientists engaged in a fierce debate about how to prevent hundreds of babies from dying suddenly and inexplicably each year. In 1985, a groundbreaking study had revealed that sudden infant death syndrome (SIDS) was rare in Hong Kong, where cultural norms had babies sleeping almost exclusively on their backs. In 1987, the Netherlands had started a public health campaign urging parents to do the same with their infants, which had resulted in a steep decline in SIDS cases, and the UK was considering following suit. Some American researchers felt strongly that the United States should do the same. Others felt just as strongly that it should not.

The reason for their opposition? The evidence was limited, and no one in the research community had been able to determine *why* putting babies to sleep on their stomachs made SIDS more likely, despite a flood of funding and an influx of research aimed at figuring it out.

"Some scientists felt it was irresponsible not to wait on launching a public health campaign until we knew the exact mechanism by which stomach sleeping causes SIDS," explained Mark Miller, coinvestigator with the Center for Integrative Research on Childhood Leukemia and the Environment at UC Berkeley and associate clinical professor and codirector of the Western States Pediatric Environmental Health Specialty Unit at the University of California San Francisco (UCSF). "Thankfully though, the 'Back to Sleep' camp won, and in the decades since the campaign was launched in 1994, the United States has seen a reduction in SIDS of about 50%." More than twenty-five years later, scientists still don't completely understand why stomach sleeping makes SIDS more likely.

Similarly, some scientists today think it's important to wait until we know exactly how, at a cellular level, exposure to certain chemicals causes cancer before we focus resources on preventing widespread exposure to them. But we have much more evidence about environmental risk factors for childhood leukemia now than we had on SIDS when Back to Sleep was launched, and Miller, who specializes in childhood leukemia research, believes there's a lesson to be learned there.

"Once we have identified a risk factor for cancer, do we really need to wait decades for all of the mechanisms to be fully understood before we take action?" he asked. "We know that fewer exposures to risk factors mean fewer cancer cases, and we also know we should be doing these things anyway—pesticides, air pollution, and carcinogenic chemicals in personal care products have all kinds of other risks too, especially for children, including asthma, hormone disruption, and impaired neurodevelopment."

While many cancer experts believe the old adage holds true—that an ounce of prevention is worth a pound of cure—that's not how we've tackled fighting cancer to date. It's estimated that only 7%–9% of all global funding goes toward prevention.[4] The rest of the billions of dollars funneled into the war on cancer goes toward pursuing cures and treatment.

While research on cures and treatments is critical, this imbalance is problematic. If the "war" was an actual military conflict, this would be like spending 93% of our resources to treat wounded soldiers and civilians, and spending only 7% on offensive or defensive measures that could prevent them from getting hurt in the first place.

Even when cancer prevention initiatives are funded and carried out, most focus on individual choices that can reduce risk, like not smoking, not drinking too much, eating healthy foods, and exercising. Those lifestyle choices certainly matter, but many, many people who make all the right lifestyle choices still get cancer, including increasing numbers of children, who generally aren't smoking or drinking. Their parents' lifestyle choices play a role, of course, but while rates of drinking and smoking during pregnancy have steadily declined, childhood cancer rates have continued to skyrocket.

Today, very little cancer funding is allocated to reducing our exposure to cancer-causing chemicals in our air, water, food, or personal care products—something that affects every single one of us, no matter how smart our lifestyle choices are.

Chemicals of Concern

New research suggests that traditional methods of screening chemicals are inadequate because they overlook the fact that humans are perpetually exposed to "acceptable" doses of many carcinogens simultaneously. They also fail to account for the chemical mixtures we're exposed to, which can interact to create even greater cancer risk than the sum of each individual chemical exposure. And traditional screening methods miss many chemicals that raise cancer risk indirectly in a number of ways, including by disrupting our natural hormonal processes.[5]

These "endocrine-disrupting chemicals" are widespread, and a growing body of research links them to hormone-related cancers. Some of these cancers, including multiple myeloma and testicular cancer, have risen dramatically since the US started tracking national cancer rates in 1975, even as other types of cancer have declined in adults.[6] Endocrine disruptors' association with breast cancer risk has been particularly well documented, and researchers have found that women bear a disproportionate burden of chemical exposure. A 2021 study identified almost 300 endocrine-disrupting chemicals in everything from hair dye, lipstick, and lotion to food additives that can increase levels of breast cancer–contributing hormones.[7] Endocrine-disrupting chemicals also contribute to obesity, which is cited as a primary risk factor for cancer.[8]

Even more troubling, when it comes to endocrine-disrupting chemicals, our exposures aren't just our own—they can increase cancer risk over multiple generations. For example, a 2021 study linked exposure to DDT, an endocrine-disrupting pesticide, to obesity and early menstruation, both breast cancer risk factors, for at least two generations. DDT was banned in the US in 1972, but this means your mother's exposures before the ban could increase not only her breast cancer risk, but also yours and your daughters'.[9] Diethylstilbestrol (DES) is another example. The endocrine-disrupting drug was widely and ineffectively used to prevent miscarriages from 1940 to 1971, but DES was banned after scientists learned that women who took it during pregnancy had daughters with a significantly higher risk of developing rare vaginal and cervical cancers.[10]

We all encounter cancer-causing chemicals in our everyday lives, but research shows that some communities face higher exposures than others. Low-income communities tend to experience higher levels of air pollution than wealthier ones, for example, but research has also shown that across all income levels, Americans of color are exposed to higher levels of harmful air pollution than their white counterparts.[11] These disparities, which are the result of decades of segregation and racist housing policies, are further exacerbated by other forms of racial and environmental injustice. For example, personal care products marketed to Black and Hispanic communities contain higher levels of endocrine-disrupting chemicals than products marketed to white communities. And because of the inequities in the US medical system, people of color often have worse cancer outcomes—for instance, Black women are 41% more likely to die from breast cancer than white women.

A New Story

I first learned Madelina's story and started researching environmental causes of cancer while working as the Pittsburgh correspondent for *Environmental Health News*, a national nonprofit news outlet focused on how changes in the environment affect human health. I'd taken the job because the organization's mission resonated with me on a personal level.

My own family has had its share of health problems. My dad, who grew up in Pittsburgh, survived a benign but difficult-to-remove brain tumor a decade ago. My younger sister, who has lived in Pittsburgh for nearly two decades, is in remission from thyroid cancer after being diagnosed at the age of twenty-five. My husband, who grew up in the region, has a rare autoimmune disorder, and his family has also struggled with various diseases, including cancer. I couldn't help but wonder whether all of their illnesses might have a common thread—especially given the area's industrial past and ongoing problems with air pollution.

Through the course of my reporting, I learned that while it's not yet possible to determine the exact causes of one person's cancer, scientists have identified eight "cancer hallmarks"—or specific changes to cells—that need to happen for cancer to develop.[12] Each of these hallmarks happens through complex interactions between genetic and external factors, including exposure to carcinogenic chemicals.

I also learned that cancer isn't typically caused by a single event or environmental exposure, but a unique set of conditions and events that produces cancer in an individual person. One emerging theory is that cancer risk is like a pie chart with a different cancer cause in each slice—things like inherited genetics and gene mutations, your parents' and grandparents' chemical exposures, and your various potential exposures to things like cigarette smoke, air pollution, or cancer-causing chemicals in food or drinking water. Everyone's pie looks a little different, but if just one piece goes missing, cancer won't develop in that person.

This made me wonder—are there slices of my loved ones' pie charts that wouldn't be there if their childhood zip codes had been different? If they hadn't lived in the shadow of steel mills and coal-fired power plants, would they have remained healthy? I don't smoke, and I try to eat healthily and exercise, but what about factors beyond my control? What slices had started filling in since I'd moved to Pittsburgh, or while I was living in Taiwan, or New York, or San Francisco?

As I continued learning about all the ways Americans are exposed to cancer-causing chemicals in our air, food, water, and personal care products, I found myself frequently overwhelmed by the enormity of the problem. Yes, I could take some steps to reduce my exposure—shopping organic when I could, paying attention to the ingredients in my shampoo and lotion and makeup, using a water filter—but I couldn't wall myself off from most chemicals. It was a realization that prompted feelings of powerlessness.

But then I discovered a national network of brilliant people working diligently to solve this problem. They are researchers, scientists, physicians, policy wonks, and concerned parents. Their numbers are still small, but their efforts are ambitious, broad in scale, and focused on systemic change—because we as individuals cannot simply shop our way out of a problem this pervasive.

Many of these efforts are being undertaken by a national coalition dubbed the Cancer Free Economy Network, which has the lofty goal of drastically reducing cancer rates by shifting the economy away from the widespread use of carcinogenic chemicals and toward safe alternatives. Since its founding in 2014, more than sixty groups and individuals have joined the Cancer Free Economy Network, including a host of health advocacy, academic, labor, environmental, and business groups, along with prominent physicians, entrepreneurs, and policy experts.

One of the network's prime directives is changing the way Americans think and talk about cancer. Progress on prevention has been slowed by pervasive cultural myths about cancer, like the notion that cancer risk is only about genetics or luck ("My grandma smoked three packs a day for forty years and never got sick."), or the idea that "everything causes cancer" so it's pointless to worry about trying to minimize exposure. In truth, it's well proven that reducing a community's exposure to carcinogens consistently reduces its number of cancer cases over time, and that focusing on prevention in ways that go beyond personal lifestyle choices can save many lives. There are also researchers, doctors, and advocates working on this issue separately from the Cancer Free Economy Network. Collectively, they form a national movement aimed at waging a new war on cancer—one that prioritizes offense and defense in addition to improving treatments for those wounded by the disease, with an emphasis on racial and environmental justice.

In the course of my reporting, I've learned the stories of both people battling cancer and people leading this movement, some of which I'm sharing in these pages. These stories aren't just about treatments or research, they're also personal histories: the familial threads and formative memories that created the reserves from which these unlikely heroes draw their resilience and ambition. Getting to know these people has given me strength to go beyond worrying about my personal consumption habits to join others pushing for change that will make the world healthier for all of us. I hope this book will do the same for you.

CHAPTER 1

Laurel: Safer Nourishment through Science

On an eighty-degree day that followed a week of snow in late April 2021, I met Berry Breene at a coffee shop in Pittsburgh. She came directly from a hike in a nearby park. Her tank top and neon orange sports bra left the port in her chest exposed, a knotty lump above her left breast. Her frame was slight, and she wore no makeup, aside from delicately penciled-in eyebrows.

We sat on the café's shaded patio, sipping cold drinks and marveling over how the sudden shift in weather had changed everyone's mood. People grinned and greeted strangers in a kind of collective exuberance about swapping out coats and boots for shorts and summer dresses overnight and finally feeling the warmth of the sun for the first time after a long, pandemic-addled winter.

A journalist colleague who was friends with Berry had connected us, knowing I was writing a book about environmental exposures and cancer. He thought I should hear her story. It was our first meeting, and Berry seemed a little nervous, which she compensated for by talking. I quickly learned that she grew up in Oil City, about 93 miles northeast of Pittsburgh, that "Berry" was her grandmother's maiden name, that she occasionally meets other women named "Bari" but had met only one other "Berry," and that strangely they had both worked at the same grocery co-op nearby. She's an artist—a painter and a musician—and works as a realtor to supplement her income. She has a half-sister who's twelve years older than she is and a brother who's fifteen months older. She lived at the top of one of Pittsburgh's many hills with a roommate she described as a "chill hippie"—meditating, sober, liked to cook. We discovered that we're just four days apart in age, both born in March 1984. And she told me that this March, just before her thirty-seventh birthday, she'd had surgery to remove a cancerous tumor from her right breast, and then she'd lost 20 pounds and most of her hair while undergoing chemotherapy.

Between sips of tea she showed me the scar from her lumpectomy—a thin, red line beneath her right armpit. She also showed me her new hair, taking off her hat and running her fingers over the short brown and gray fuzz that coated her scalp.

———

In September 1993, more than 500 people gathered in downtown Boston for the city's third annual Massachusetts Breast Cancer Coalition march. After the walk, event organizers asked anyone who'd experienced breast cancer to step inside a large outline of the state of Massachusetts they'd painted on the ground. Suddenly the shape was crowded with bodies, bandanas or headscarves covering heads that were bald from chemotherapy. People cried and clung to their spouses or children.

"With all of these pink ribbons everywhere, it's hard to believe now that not long ago breast cancer was a hidden disease," Julia Brody, the executive director of Silent Spring Institute, a breast cancer prevention research nonprofit, told me in 2020. "In the 1990s, women were just beginning to talk about their experiences publicly and connect with other women who'd had the disease. For many of the women at that rally, this was the first time they'd ever publicly acknowledged that they'd had breast cancer. It was incredibly emotional seeing that they weren't alone."

At the time, the mortality rate from breast cancer was around 31 deaths per 100,000 people in the US—significantly higher than the current mortality rate of 20 per 100,000. The National Breast Cancer Coalition, including the Massachusetts chapter that sponsored the march, circulated a petition urging President Bill Clinton to develop a national strategy to address the disease.

"I take care of women with breast cancer every day and I have seen too many patients die," Dr. Susan Love, one of the founders of the National Breast Cancer Coalition, told the *Boston Globe* reporter who covered the march in 1993. "We have to be so obnoxious that they cure this disease just to get rid of us." Their plea was effective. Over the next decade, Congress allocated hundreds of millions of dollars for breast cancer research, and mortality rates dropped by about 24%.

But a quieter, parallel movement was also being launched at the same time—one aimed at stopping breast cancer before it started.

The Massachusetts Breast Cancer Coalition had been active since 1991, but 1993 was the first year the state had ever published cancer rates by town. The report revealed that nine towns on Cape Cod had significantly higher rates of breast cancer than the state average. The group called for a scientific investigation into why, but they were frustrated by the lack of existing laboratories capable of performing the investigation. So with a few million dollars in funding from the Massachusetts legislature, they founded their own, bringing in scientists with diverse expertise. They named it Silent Spring Institute in honor of Rachel Carson, whose groundbreaking 1962 book *Silent Spring* brought national attention to the dangers posed by many pesticides, effectively launching the modern environmental movement. Carson died of breast cancer two years after the book came out.

"They were really a remarkable, visionary group of women," Julia said, referring to the activists who founded Silent Spring Institute. "Now, they say they were naive at the time and didn't realize how hard that kind of investigation was going to be, but that's lucky—if they had, they never would have achieved so much."

When the researchers first went to work trying to figure out why Cape Cod had elevated levels of breast cancer, they quickly discovered the difficulty of such an undertaking. Cancer is a complex disease with many causes, and while scientists generally know how cancer occurs at a cellular level, it's still virtually impossible to trace one person's cancer to one or more specific environmental exposures—let alone tracing exposures for numerous people, all with different health histories and habits. In 1994, only a limited number of studies had set out to do such a thing, so Silent Spring Institute researchers had to develop methodologies as they went.

They enlisted help from researchers at Boston University, Tufts, Brown University, and Harvard and conducted extensive analyses of the cancer data available at the time. They also developed new scientific methods to test air, dust, and urine for chemicals linked to breast cancer risk, then used those methods to test samples from 120 households on Cape Cod. They tested wastewater, groundwater, and drinking water for potentially cancer-causing compounds, and conducted interviews with 2,100 Cape Cod women, both with and without histories of breast cancer, to collect decades' worth of data on risk factors, health histories, habits and activities, and places of residence. The project was so thorough and innovative, according to Brody, that it became a model for environmental health studies across the country.

In the end, the researchers didn't find a smoking gun that explained the elevated cancer rates on Cape Cod, but they did find a surprising number and variety of potentially cancer-causing chemicals in most of the homes they looked at. Some, but not all, of these chemicals showed up at higher levels in Cape Cod homes than had been found in studies done in other places, including the pesticides DDT, chlordane, carbaryl, methoxychlor, pentachlorophenol, and propoxur, several of which are linked to higher cancer risk. They theorized that the elevated levels of these pesticides in Cape Cod homes could have come from unique local farming practices (cranberry bogs are prevalent in the region) or aerial spraying to control mosquitoes and gypsy moths. Those pesticides might have contributed to the region's elevated cancer rates, but it was hard to know for certain.

While the study didn't exactly answer the riddle that Silent Spring Institute had set out to solve, the researchers quickly realized their data had much larger implications: Many of the chemicals they found in people's homes came from drinking water contamination and common household products, so it seemed likely that homes across the country were experiencing similar exposures—but for the most part, no one else was looking for them, at least in the US. They had inadvertently discovered that Americans were likely being exposed to a host of potentially cancer-causing chemicals on a daily basis, and were largely unaware.

Once the researchers recognized the scope of the problem, they realized that their work would have greater impact if they focused on helping people avoid these exposures in the first place—so they pivoted from studies that looked backward to ones that looked forward. Today, Silent Spring Institute is staffed by twenty-two people, most of them women, focused on identifying chemicals that could cause cancer and helping people avoid them.

This kind of work requires constant innovation. Current tests to determine whether a chemical causes cancer often require lab animals and are slow and expensive, so Silent Spring Institute scientists are developing new methods to quickly screen thousands of chemicals and identify which are likely to increase breast cancer risk. The organization communicates these findings to the public so we can pressure industries and policymakers to keep harmful chemicals out of food and personal care products. The organization also wants to expand the definition of a carcinogen, moving beyond just chemicals that damage DNA to include things like endocrine disruptors, chemicals that stimulate inflammation, and chemicals that affect the body's immune responses, all of which have been found to raise cancer risk.

"There is an array of pathways that affect cancer," said Julia, "but many chemicals on the market have never been tested for these health effects."

––––––––

On a crisp fall evening in November 2020, Laurel Schaider, a senior scientist at Silent Spring Institute, put a laptop at an empty place at her dining room table so I could virtually join her family for dinner. Laurel, her husband, Grant, and their children, Ethan and Elena, had ordered from their favorite taqueria in their Boston suburb, and I had ordered from my favorite burrito spot in Pittsburgh. In the midst of a global pandemic, this was the closest we could get to safely having dinner together. As an environmental chemist and public health researcher, Laurel wasn't taking any chances, but COVID-19 wasn't our topic of discussion. Instead, we talked about whether our dinners contained cancer-causing chemicals.

"These come wrapped in aluminum foil, so I'm generally not worried about the wrappers, at least," Laurel said between bites of a plantain burrito. Laurel has wavy brown hair parted down the middle, pale skin, and blue eyes, and she wore oval glasses with purple frames. She and I first met in 2019 when I interviewed her for a story about potential carcinogens in tap water. Two years before that, Laurel had led a groundbreaking study, which found that harmful chemicals with the potential to increase cancer risk are common in fast food and takeout packaging, including the grease-proof paper that's placed under pizzas and wrapped around everything from burgers to baked goods.

Of the 400 wrappers and containers she and her colleagues collected and analyzed from various restaurants, 33% contained possible carcinogens. The study generated a lot of buzz at the time, so Laurel conducted a follow-up study in 2019, which confirmed that people who frequently eat takeout, fast food, and pizza have higher levels of those chemicals in their bodies than people who regularly cook at home.

The chemicals Laurel studies in food and water are per- and polyfluoroalkyl substances, also referred to as PFAS. Dubbed "forever chemicals," PFAS are a class of more than 12,000 humanmade fluorinated compounds that were invented by DuPont in the 1930s. In addition to being sprayed on food wrappers and takeout containers to make them grease resistant, they've also been used in products like stain- and water-resistant clothing, nonstick pots and pans, and "spill-proof" carpets and furniture. DuPont (the company responsible for Teflon, Mylar, and Kevlar) and 3M (the company that invented Scotchgard, Scotch Tape, and Post-it Notes) started mass producing PFAS in the 1950s, and their use has become widespread across the globe. PFAS don't break down naturally, so they can accumulate in the human body over time. Most PFAS haven't been thoroughly studied for health effects, but those that have been studied are linked to various health issues, including decreased birth weights, thyroid disease, decreased sperm quality, high cholesterol, pregnancy-induced hypertension, asthma, ulcerative colitis, and testicular and kidney cancer. Some evidence also suggests that certain PFAS could increase the risk of breast cancer.

Laurel's 2019 study on takeout was the first to link consumption of to-go foods with levels of PFAS exposure in Americans.[1] Laurel and her colleagues analyzed data collected between 2003 and 2014 from more than 10,000 participants in the National Health and Nutrition Examination Survey—a program of the US Centers for Disease Control and Prevention (CDC) that tracks dietary, health, and nutritional trends in the United States. The survey asks detailed questions about participants' diets, documenting what they ate over the last 24 hours, 7 days, 30 days, and 12 months. Participants also provide blood samples that are analyzed for certain chemicals, including a number of different PFAS. "We found that every 100 calories of food purchased at a grocery store and prepared at home rather than being purchased from a restaurant was associated with 0.3% to 0.5% lower levels of PFAS in our bodies," Laurel explained.

High levels of PFAS have previously been detected in microwave popcorn bags (the insides of the bags are sprayed with the chemicals to keep buttery grease from seeping out), and Laurel's study also found that people who had eaten microwaved popcorn daily over the past year had around 39% and 63% higher levels of two kinds of PFAS in their bodies, respectively, than people who had not. Her findings provided some of the clearest evidence yet that these chemicals are migrating from food wrappers into our bodies.

Even knowing this, Laurel doesn't completely avoid ordering take-out. Since the start of the pandemic, getting dinner from their favorite taqueria or Chinese restaurant once or twice a week had become a family tradition—something to look forward to during a stressful time. Like many families, all four of them spent most of 2020 in the same house together, everyone doing work or school at their respective screens.

Twelve-year-old Ethan is polite and a bit reserved, "Like us," Laurel said, referring to herself and Grant, "while Elena is the family extrovert."

Eight-year-old Elena has bright brown eyes and an abundance of energy, and is desperately trying to convince her parents to get a dog. Both kids vaguely know what their mom does for a living. Between bites of burrito they proudly showed me pictures they'd been drawing earlier in the day—basketballs, desert islands, and, of course, dogs—and explained that their mom tries to help people by protecting them from chemicals that can hurt them. Elena repeated the word "PFAAAAS," drawing out the A for a goofy effect over and over, but she shrugged and went back to drawing when I asked if she knew what PFAS are.

"I admire Laurel's work very much," said Grant, who has dark hair and glasses and does marketing for a cloud software company. "My world is very profit focused, so I think her community-focused work is really admirable." Grant and Laurel are both middle-aged, but they still look almost as young as the day they met as undergraduate students at the Massachusetts Institute of Technology, where they bonded over the sudden joy of being self-described "nerds" who'd found themselves in the company of other nerds almost exclusively for the first time in their lives.

As we ate and talked, I considered the other potential carcinogens that could have made their way into my burrito. The flour tortilla might have been made from crops treated with pesticides, traces of which could linger even after all the processing that turns wheat into a tortilla. There have been more than 500 active ingredients used in pesticides in the US since the 1970s, many of which haven't been tested for safety. Some of the most dangerous pesticides have been banned, but a number of pesticides currently being used in the US contain carcinogens, and the US uses vastly larger quantities of dangerous pesticides than other leading agricultural nations. A 2019 study found that the US still uses seventy-two pesticides that either have been banned or are in the process of being completely phased out in the EU, seventeen that are banned or being phased out in Brazil, and eleven that are banned or being phased out in China.[2] The same study found that in 2016, the US sprayed 322 million pounds of pesticides that are banned in the EU, 26 million pounds of pesticides banned in Brazil, and 40 million pounds of the kinds banned in China. The grilled onions and peppers in my burrito, the corn used to make my tortilla chips, and the tomatoes, peppers, and tomatillos used to make the salsa and spicy hot sauces I drowned it all in were also likely grown using pesticides.

Meanwhile, the refried beans in my burrito might have come from a can lined with bisphenol A (BPA). The compound is still widely used in can and bottle linings to stop food spoilage and prevent corrosion. It's also used in hard, sturdy plastics, and to make receipt paper responsive to thermal printing. BPA, which is an endocrine-disrupting compound, has been found to migrate from these objects into our bodies. These chemicals mimic our natural hormones, which can disrupt everything from metabolism and sleep to normal growth patterns in children and the ability to reproduce in adults. These effects are also thought to play a role in the development of hormone-associated cancers, such as breast and ovarian cancers[3] and prostate cancer.[4] Even more troubling, many products labeled "BPA free" use replacement chemicals that are molecularly similar to—and every bit as dangerous as—BPA.

Well aware of all these hazards, Laurel makes an effort to cook at home and shop organic when she can. Grant admittedly worries less about such things, but regardless of who's shopping, Laurel knows there's no amount of vigilance that could completely protect her kids from exposure to potential carcinogens in their food.

In recent years, scientists have detected PFAS in all kinds of foods, far beyond just fast food and microwave popcorn—from dairy milk at family farms to grocery store items like beef, fish, and produce. Food packaging can be a source of PFAS getting into foods, but the plants and animals we eat can also absorb PFAS from the environments they're raised in. In those cases, one likely source of PFAS contamination is biosolids, a commonly used type of fertilizer. Biosolids are essentially chemically cleaned waste that includes human feces. They're made by removing solid waste from raw sewage sludge and treating it to filter heavy metals, adjust pH levels, and remove harmful bacteria, viruses, and parasites. PFAS can get into biosolids from industrial waste or human excrement that winds up in sewage, and most states don't test or treat the fertilizer for PFAS. Biosolids are generally cheaper than chemical fertilizers and are sometimes given to farmers for free. In some states, farmers are even paid to apply biosolids to their fields to help wastewater treatment facilities get rid of the stuff. Produce plants grown in soil contaminated with PFAS uptake the chemicals into their roots, leaves, and fruit. Biosolids can't be used on organic crops, but if the soil where organic produce is grown has ever been used to grow conventional produce—even many years ago—it's possible that PFAS could still linger in the soil and wind up in organic produce.

"This isn't a problem we can just shop our way out of," Laurel said. "Not everyone can afford to buy organic food, which creates inequities in these harmful exposures, and even for people who can, there are just too many chemicals to worry about, and things aren't always clearly labeled or disclosed."

———

Berry lives in a red brick house at the end of a quiet, dead-end street in Pittsburgh's Troy Hill neighborhood, which sits atop a steep hill overlooking downtown and the Allegheny River with its many yellow bridges. It's an old fixer-upper she's been making her own since buying it a few years ago. The views from the windows at the back of the house in the summertime are breathtaking—big sky over rolling green hills peppered with row houses.

There's a small flower garden in front of the teal gate that leads to her front porch, and nestled among the plants is a sign that reads, "We believe Black lives matter, women's rights are human rights, love is love, no human being is illegal, science is real, kindness is everything."

The interior of the house is painted white with bright blue and turquoise accents, and decorated with quirky antiques and vintage furniture: velour armchairs with kitschy crocheted pillows, mismatched lamps, a curio cabinet filled with old family photos and knickknacks, and a pew from a now-demolished church that serves as a bench. There are house plants everywhere, in antique brass pots and chalices, sprouting from golden bowls, and dangling from the ceiling in macramé hangers. The walls are decorated with thrifted artwork, Berry's photos in vintage frames, and a large tapestry she confiscated from her grandmother's house. In the hall at the top of the creaky staircase, there's a triptych of vintage black and white photos of nude women posing artfully.

An acoustic guitar rests on a stand beside the fireplace. Berry's dad introduced her to Elvis, the Beatles, Janis Joplin, Motown, Doowop, showtunes, and '50s and '60s rock when she was a kid. She'd always loved singing but was too shy to perform in public until her dad died unexpectedly in a car accident in 2015. The two were close. She channeled her grief into making music, but the thought of performing publicly still made her weak. She pushed herself to do it anyway. "Once he passed," she said, "I realized life's too short to be afraid of things you enjoy."

The first time she performed with the band in a bar she froze, unable to force any sound from her throat, despite countless hours of rehearsals. "That moment felt like an eternity," she said. She pushed through anyway, croaking out the first few words to the song until her full voice returned to her. "Afterward," she said, "I realized no one else in the room had even noticed that brief moment of silence and panic."

Above her guitar by the fireplace, an abundance of books filled the shelves on either side of the mantle: art books, literary classics, new novels, a birding book, some yoga books. And a copy of Rachel Carson's *Silent Spring*.

———

Silent Spring Institute is housed in a building in a Boston suburb that once manufactured ship rigging and trolley pulls using steam power. It's full of tall windows, exposed brick and bright white walls, wooden beams overhead, and the occasional Pilates ball serving as an office chair.

The building doesn't house a physical laboratory. Instead, the researchers partner with academic and commercial labs, which gives them access to the most advanced techniques available.

For example, they recently partnered with a group of nuclear physicists at the University of Notre Dame who'd figured out that they could adopt a technique used to identify fluorine in bones and rocks to look for fluorinated chemicals—PFAS—in other materials, like fast-food packaging. The technique is called particle-induced gamma-ray emission spectroscopy (or PIGE, pronounced "piggy"). "They were the first to put these two pieces together and figure out that PIGE could be used to rapidly screen for PFAS," Laurel explained. "You need a particle accelerator, so only specialized labs can do it."

Silent Spring Institute partnered with the researchers at Notre Dame, the US Environmental Protection Agency (EPA), the Washington State Department of Ecology, the Environmental Working Group (a research and advocacy group), and the Green Science Policy Institute to apply PIGE to PFAS research. Laurel's 2017 paper on PFAS in fast-food packaging came out of this collaboration.

According to Julia, the need for this kind of research is urgent. "There's so much talk about breast cancer now, but very little of it is about prevention, and even less takes environmental factors into account," she said. "You really shouldn't have to have a chemistry degree to be able to safely go grocery shopping and buy personal care products. It's difficult to know what's in things even for those who *do* have chemistry degrees."

She described a 2017 project where Silent Spring Institute invited people to have their urine tested for ten endocrine-disrupting chemicals. About 800 people volunteered, including *New York Times* columnist Nicholas Kristof, who wrote about the experience.[5] The researchers found that consumers who actively read labels and tried to avoid certain chemicals had slightly better test results than those who didn't think about it at all—but that those efforts still didn't fully protect them.

"Even our own scientific staff who participated couldn't eliminate their exposure to some of these chemicals—especially BPA and related compounds," Julia said. "No matter how vigilant you are, it's just not possible to avoid them entirely. Even if you could control what's in the products you buy, we're all still exposed to carcinogens in air pollution and drinking water. Individuals can do some things to help, but none of it is going to be as effective as policy change."

As an example of policy change that has worked, Julia pointed to the US banning leaded gasoline in 1996, which contributed to a more than 90% decrease in average blood lead levels among Americans over a forty-year period.[6] This change is credited in part for a 4- to 5-point increase in the average American IQ.[7]

Laurel is one of just a few researchers at Silent Spring Institute who focuses almost entirely on one group of chemicals—something that grew out of her discovery that exposure to PFAS is so widespread and pervasive. "When people are shopping at the grocery store or ordering takeout, they generally just assume that food is safe," she said. "People don't usually think about whether there are dangerous chemicals on their food wrappers or even in the food itself—and rightfully so. They shouldn't have to. That's really what I'm working toward."

Laurel's office at Silent Spring Institute is decorated with pictures that Elena and Ethan have drawn over the years. The building was mostly empty due to COVID-19 when she gave me a virtual tour in the fall of 2020. She and her colleagues had mostly worked from home since the previous March, but Laurel popped in every now and then to water the office's many plants and move them to the sunniest spots on unused desks and conference tables.

The wall calendar in Laurel's office was a month behind, so she flipped the page, saying, "I do try to change the calendar when I come in so it doesn't feel quite so much like the apocalypse."

There was one framed picture on the wall—a photograph of a

wooden rocking chair on the porch of a summer cabin facing a beautiful landscape, given to Laurel by a family she worked with tracking chemical exposures in their community. The photo serves as a reminder that the daily grind of her work—which now often includes things like project management and budget oversight—is ultimately aimed at helping real people.

Laurel has been interested in the environment for as long as she can remember. She was born in 1975 and grew up in a town that's considered the birthplace of the modern conservation movement—Concord, Massachusetts, home to Walden Pond, the subject of Henry David Thoreau's book *Walden*. She was an only child who liked reading, math, puzzles, knitting, and walking in Thoreau's beloved woods. When she was a freshman in high school in 1990, the twentieth anniversary of Earth Day dominated headlines.

She became concerned about climate change and the fate of the rainforests and grew obsessed with recycling, helping her school launch its paper recycling program and driving her parents crazy by constantly picking things out of the trash.

Laurel knows now that she was lucky to have teachers who encouraged her interest in science and math at a time when many girls were discouraged from those interests—even if she didn't appreciate it at the time. "I was on the math team, which did not up my cool factor at all," she said. The retired math teacher who coached the team would learn the students' schedules, stop by their classrooms, and bring them practice questions, which Laurel found highly embarrassing. "Everyone could see I liked the math part, but I didn't need to broadcast that I was even nerdier than everyone already thought I was," she said, laughing. "But in retrospect, I'm really grateful that he encouraged me like he did."

When Laurel started college at the Massachusetts Institute of Technology, women represented only about a third of the student body, but the tide was changing. By the time she graduated, it was almost fifty-fifty. Laurel chose environmental engineering as her major and spent a summer helping grad students study historical patterns of nutrients in soil at Harvard Forest, a 3,000-acre ecological research site in central Massachusetts. She loved the work, and it made her more certain that she wanted a career as a scientist. She went on to get a PhD in environmental engineering from UC Berkeley, then did her postdoctoral research studying heavy metal exposure at an abandoned mining site in rural Oklahoma through the Harvard T.H. Chan School of Public Health. "That's when I realized I was interested not just in the environment out there, but in how these pollutants in the environment affect our bodies and our health," she said.

In Oklahoma, Laurel found herself in a totally alien landscape: Flat, wide-open agricultural fields punctuated by hills of sand and rocks made of mining waste—some big enough to drive up—that were riddled with cadmium, zinc, and lead. She and her colleagues analyzed plants, soil, and house dust and found that the mining waste was contributing to lead poisoning in kids belonging to Native American tribes in the region. "Before this I'd mostly been in a lab," Laurel said. "This was my first time working with a community group, and I found it really rewarding. If we found high levels of lead in a plant with cultural significance to one of these tribes, it wasn't just a theoretical thing for them—it had actual implications for their health. The stakes were high because people really needed the answers to these questions we were trying to answer."

After four years doing postdoctoral work through Harvard, Laurel learned about Silent Spring Institute through a colleague, and the organization's mission immediately resonated with her. Like most Americans, she knew and loved people who'd been affected by cancer. Cancer prevention felt like a noble mission, and working there also presented an opportunity to continue working with communities. "I appreciated Silent Spring's focus on educating the public and policymakers to promote stricter regulations," Laurel said. "We're not an advocacy group, but we're also not content to just publish scientific research, talk about it with other scientists, and move on."

————

While some of our dietary choices can reduce our exposure to potential carcinogens in food, it's difficult to avoid chemicals that come out of the faucets in our homes. This is especially true of PFAS. They're largely unregulated, which means most water authorities don't regularly test for them, let alone take steps to remove them. PFAS have gotten into water supplies throughout the country from industrial waste and a specific type of firefighting foam used in training drills at military bases and airports. For decades, there were no specific cleanup protocols in place for the firefighting foam, so it was simply washed away, putting PFAS into the surrounding soil and groundwater.

National data on PFAS in drinking water are still lacking, but in 2020 the Environmental Working Group tested drinking water samples from forty-four places in thirty-two states. They found only one location that had no detectable PFAS, and only two other locations with PFAS levels below the threshold that poses health risks. They estimated that at least 200 million, or about 61% of Americans, have been exposed to PFAS through their drinking water.[8]

In 2019 the same group analyzed water testing reports from more than 48,000 water utilities in all fifty states for other harmful chemicals and found twenty-two carcinogens that are prevalent in American drinking water, which they estimate cause more than 100,000 cases of cancer across the country. The list includes arsenic, uranium, and benzene, along with a long list of disinfection by-products—things like trihalomethanes, chloroform, and bromate—which are formed when natural compounds in source water combine with the chlorine used to disinfect drinking water.

Many federal drinking water limits for harmful contaminants haven't been updated in twenty to forty years. Regulators consider not only the health effects of contaminants, but also the cost of removing them from drinking water, and limitations of the filtration technology available at the time laws are implemented. Consider how much other technologies have evolved in that period—twenty years ago we didn't have smartphones, YouTube, or Netflix, and forty years ago virtually no US homes had internet access—and you'll have a sense of how much better and cheaper large-scale water filtration systems could have gotten in that time span. As a result, even drinking water that meets all federal standards in the US can contain harmful contaminants. Using home filters or buying bottled water can help, but not everyone can afford these options.

Laurel leads Silent Spring Institute in two large projects related to PFAS in drinking water. One is working with the Agency for Toxic Substances and Disease Registry (ATSDR) to track exposures in communities contaminated with PFAS; the other is a five-year project funded by the National Institute of Environmental Health Sciences that researches the impacts of PFAS exposure on children's immune systems.

Numerous studies have looked at exposures in communities with PFAS-contaminated water, but the ATSDR study is the first time it's been attempted at the national level. To do this, the CDC funded seven projects in PFAS-exposed communities across the country to be led by scientists like Laurel. Researchers at each site are required to use the same techniques for gathering data.

"I don't know of any other study that's been done like this or at this scale," Laurel said. "All the teams have what's called a collaborative agreement—everyone has to do exactly the same thing the same way, then the teams can also opt to do additional testing or analysis. We're recruiting participants in these communities to submit blood and urine samples and administering a questionnaire about where they live, the water they drink, and their occupations and health history. A lot of these communities aren't large enough to study alone, so looking at multiple sites and being able to combine the data will give ATSDR more statistical power to look at associations for PFAS exposure and health." She added that the communities also have varying degrees of PFAS exposure, so researchers hope to tease out comparisons between them.

Emerging research indicates that in addition to causing certain types of cancer and other diseases, PFAS also damage the immune system and render vaccines less effective. PFAS-exposed children have shown decreased responses to common childhood vaccines through their teenage years, and studies of exposed adults have shown lower than average responses to flu vaccines. One 2020 study linked higher levels of PFAS exposure to more severe symptoms of COVID-19.[9] Laurel and six other researchers coauthored a 2020 op-ed about PFAS potentially putting people at greater risk for getting severe COVID-19 and making the vaccine less effective, urging the US government to begin regulating the chemicals in drinking water at the federal level.[10] These projects aren't explicitly tied to cancer prevention, but the functioning of our immune systems also plays an important role in the development and treatment of cancer.

———

In early June 2020, with the COVID-19 pandemic in full swing, Berry noticed that when her guitar was pressed against her body, she felt a strange pressure on the underside of her right breast. By then, Berry was regularly performing as a solo singer-songwriter, and she sang in an indie folk band that had been on tour and released an album. She still occasionally got jitters before going on stage, but when the pandemic forced her band to cancel all their live gigs, she missed the high that came from performing. She still turned to her guitar for solace.

Berry is small, and the spot on her breast where she felt pressure was often in contact with her instrument, so she spent a month trying to convince herself that it was some sort of callous. But in mid-July she asked her boyfriend whether he thought it had gotten bigger since the last time she'd asked him to feel it a month before. He said yes, without hesitation.

"That was when I finally admitted to myself that internal calluses aren't a thing," she said. "It was a lump."

————

While learning Laurel's story over burritos, I wondered how she avoids paralyzing fear every time she feeds her kids. I also reconsidered whether this was a rabbit hole I really wanted to continue digging into. Wouldn't it feel better to choose ignorant bliss over learning how the things I loved—eating food and drinking water both being high on the list— were raising my cancer risk?

But there's good news: Laurel and her colleagues at Silent Spring Institute are far from being the only people diligently working to get carcinogens out of our food and water. "It's very rare for a single study to immediately change the market or change policy in a meaningful way," Laurel said, "but I've seen my work contribute to meaningful changes over time that have helped protect a lot of people."

Organizations doing similar work in the US include the Environmental Working Group, the Environmental Protection Network, the Cancer Free Economy Network, the Environmental Justice Health Alliance for Chemical Policy Reform, Coming Clean, and Safer Chemicals Healthy Families, among others. These types of organizations increasingly partner with one another and rely on collaboration, and their collective efforts are paying off.

Awareness about the dangers of PFAS is increasing among US lawmakers. Numerous bills related to protecting Americans from these chemicals have been proposed in recent years, including 2021's Keep Food Containers Safe from PFAS Act, which would ban PFAS in food packaging in the US if passed. Denmark banned PFAS in food packaging in 2020, and a handful of states have followed suit, including Colorado, Maine, New York, Vermont, and Washington. Thanks to a combination of those bans and consumer pressure, several large restaurant chains have also committed to phasing out PFAS in consumer-facing food packaging, including Chipotle, Wendy's, Panera, Taco Bell, and McDonald's. A handful of grocery chains, including Whole Foods, Trader Joe's, and Albertsons, have also committed to eliminating food packaging containing PFAS. And in the fall of 2021, likely in response to mounting public pressure, the US Environmental Protection Agency also published a new timeline for various steps to regulate, clean up, and research PFAS. Environmental groups have said the plan doesn't go far enough or move fast enough, but it's a start. In the meantime, states have begun setting limits for specific PFAS in drinking water in the absence of federal regulations (which are expected sometime in 2023), including New Jersey, Michigan, New York, New Hampshire, Massachusetts, and Vermont. Some states, like Maine and Colorado, have also passed legislation that restricts the use of PFAS in consumer products.

While regulations can take a long time to pass, progress is also being made through changes in the marketplace. Nearly every major chemical company is beginning to experiment with "green chemistry"—designing chemical products and processes that reduce or eliminate hazardous substances—with hopes of developing the next big thing and getting ahead of new regulations. Market analyses show that products marketed as being made with green chemistry consistently outperform their more toxic counterparts, and the market share of household and personal care products marketed with green chemistry is quickly increasing, growing from 10.1% of the total market for personal care and household products to 14.3% between 2015 and 2019.[11]

For example, Valspar (which was acquired by Sherwin Williams in 2017) has been working to develop new BPA alternatives—ones that have actually been proven not to cause endocrine disruption or cancer—to line food cans with, despite a lack of US regulations requiring them to do so. This is likely a result of both consumer pressure in the US and bans on the chemicals in other parts of the world. The company assumes similar bans are coming elsewhere, and they want to remain competitive in the global market. In 2009, Valspar enlisted some of the scientists who've researched BPA most extensively and criticized its continued use most vocally to help them develop safer can linings. After years of extensive testing, they found a compound that can stop food spoilage (BPA's most useful function) but that doesn't appear to disrupt normal human hormone function, and is less likely to leach into food than BPA and existing replacement chemicals. That compound is now being used in a small but growing share of food and drink cans across the US. If it proves effective and safe over time, it could eventually become the new standard and replace BPA and similarly harmful substitutes.

These might seem like incremental changes, but the sum of all these shifts will reduce cancer rates. And efforts to get carcinogens out of food and water are just the beginning.

CHAPTER 2

Ami: Safer Beauty through Racial Justice

Five-year-old Zaria bounced up and down as her mother lifted a wooden box full of cosmetics from the top of her dresser and set it on the bed. Her mom's makeup is usually off-limits, so Zaria could barely contain herself as she pulled out compacts and tubes of foundation, lipstick, and mascara, asking, "What's this?" "Where does this go?" and "How do you put this on?"

Zaria has a wide, gap-toothed grin, pierced ears, and a wardrobe full of sparkles. She's nearly fluent in Spanish—she's been in immersion classes since she was two—and she's passionate about reading and math. And makeup.

"Hey, slow down, you have to be gentle," said her mother, Ami Zota, when Zaria aggressively stuck her fingers into an eyeshadow palette, smudging the tidy rounds of powder.

As Ami (pronounced AH-me) explained what each item of makeup was and where it was applied, Zaria begged her to try them on.

"You can pick one thing to try," Ami said. Zaria chose blush. Ami tenderly swept Zaria's hair aside and brushed the powder onto her smooth, brown cheeks. "How does that feel? Does it tickle?"

"No," Zaria said, closing her eyes, "it feels nice."

Within just a few minutes, Zaria had overcome Ami's resistance and convinced her to also apply red lipstick, shiny lip gloss, and shimmery eyeshadow.

"It's very subtle," Ami said sarcastically as Zaria puckered her lips while admiring her new look in the mirror.

"What does subtle mean?" asked Zaria.

"It means it's not too obvious," Ami explained. "Remember you don't need any of this stuff, okay? You'll always look beautiful just the way you are."

Ami, a public health researcher and professor at Columbia University's Mailman School of Public Health, is a second-generation Indian American with dark, curly hair, brown eyes, and a nose ring. She doesn't wear makeup often, but when she does, she favors bold, red lipstick. Ami and her family moved to New York in 2022 but were living in Washington, D.C., at the time. I met Ami through work after she launched a program in collaboration with *Environmental Health News* dedicated to telling stories by historically underrepresented and marginalized voices in science. As she told me why she tries to avoid cosmetics that contain phthalates, parabens, metals, fragrances, and talc, Zaria found a large white banana clip and affixed it above her forehead, creating a spray of bangs that made her look like she'd walked straight out of an '80s teen movie.

"Oh my goodness, you look like you're 13," Ami said. "That's crazy. Go show your dad."

Later, while chowing down on a dinner of crispy fish and broccoli at the kitchen table, Zaria repeatedly interrupted the conversation I was having with her parents to ask, "Dad, is the lipstick still on? Can you still see it on my lips?" After one such outburst she added, "Good thing I have a mirror by my bed so I can keep checking while I'm sleeping."

Her parents laughed.

"Oh no, Honey," Ami said. "You can't sleep with makeup on—it's not good for your skin, and it would get all over your sheets. You're getting a bath tonight."

"Can we put on some more lipstick the day after tomorrow then?" she asked.

"No," Ami said firmly to Zaria, then explained to me, "Her school is closed tomorrow for the inauguration. She wants more lipstick the day after tomorrow because she's trying to impress her classmates."

It was January 19, 2021, the day before the inauguration of Joe Biden as the 46th president of the United States and Kamala Harris as the nation's first woman, first Black, and first South Asian vice president. Before trying on her mother's makeup, Zaria had proudly showed me her new copy of *Ambitious Girl*, a feminist children's book written by Kamala Harris's niece, Meena Harris. I couldn't help but notice that Zaria's life bears some resemblance to the new vice president's. Kamala Harris's mother was an Indian American scientist who researched breast cancer, while Zaria's mom conducts research aimed at preventing cancer and other diseases caused by toxic chemicals in cosmetics and personal care products.

"Pleeeease can I put on lipstick again just one more time?" Zaria wheedled. "I just want to be fancy for one day."

"No, I'm sorry," Ami said. "You know when you can be fancy? For that wedding we're going to this summer. Maybe one of your cousins can help you put makeup on for that."

"Oh wow," said Zaria's dad, Dave. "That will be such a special treat!"

Zaria frowned but nodded, then asked to be excused to go have some screen time. After she left the table Ami said, "This might become a problem. She's really into precedent setting. Now that we've let her do it once she won't stop asking."

"A lawyer in training," Dave added, laughing.

"I have no idea where she's even getting interested in this stuff," Ami said. "It's not coming from me. I think it's just so ubiquitous in our culture that she's picking it up, which is a little unsettling."

One of Ami's earliest experiences with makeup was her own mother showing her how to use kajal, or kohl—dark black eyeliner used throughout the Middle East and North Africa, the Mediterranean, and South Asia. In parts of India, new mothers traditionally applied kajal to their baby's eyes to ward off evil spirits and negative attention. Ami used kohl regularly as a teen, but later learned that it often contains heavy metals and has been recognized as an important source of harmful exposures in some immigrant communities. It's one of many parts of Ami's personal and family history that inspired the work she does today.

Ami grew up in rural North Carolina and says that as a kid, she never imagined she would be a scientist or a professor. "I barely knew these professions existed," Ami said, "and I certainly didn't see any female Indian Americans in these roles."

Ami's life has been vastly different from the lives of previous generations of women in her family. Her mother, Lata, grew up milking cows and water buffalos in a rural part of the northeast Indian state of Gujarat. Lata stopped going to school after fourth grade to stay home and help raise her seven siblings before entering an arranged marriage with Ami's father, Ramnik, who worked the nightshift at the village gas station. Ramnik wanted more opportunities than life in their hometown offered. He'd always been gifted in school, so he applied for scholarships and received one at a school in the nearest city. When he moved to the United States to complete his medical residency at a hospital in Chicago, Ami's mother stayed behind and lived with his parents until he was able to send for her the following year. When she first moved to the states, Lata didn't speak a word of English. She was still very close with her family, who were now half a world away, and she missed them. She was homesick. "Life was very, very hard for her when she first arrived," Ami said.

To quell their loneliness, Lata and Ramnik worked on convincing relatives to come settle in the states. Soon cousins and aunts and uncles started showing up one by one. Eventually, Ami's mother came to be seen as the matriarch of the entire extended family, a role that helped her thrive. Now, a generation later, Ami guesses that there are about forty people from her extended family in the US, all of whom followed her parents.

Soon after her arrival in the states, Lata gave birth to her first daughter, Ami's older sister, Sejal. The young family moved to North Carolina, where Ami was born in 1977, followed by her younger sister, Rita. All three children excelled academically.

"We've always been known as a willful, driven, intelligent trio of girls," Ami said. "Generations of our family have been entrepreneurs on a small scale, but my dad was the first doctor, and then my sisters and I were the first women in our family to pursue advanced educations."

Ami spent her early years in Raeford, North Carolina. Her elementary school was diverse, with many Black and Indigenous students, but she remembers how it felt to be the only Indian American that many of her classmates and teachers had ever encountered.

"People in Raeford didn't even know where India was on a map," she said. "I remember one time one of my friends came over to play when my grandparents were visiting from India. She got so scared of them she started crying. My first grade teacher didn't know how to pronounce my name so she called me Army. We were just really, really different from everyone there. That's hard when you're young."

"There's also some struggle about your social identity when your parents grew up in India and you're raised in America," she added. "What does that mean when you're kind of both? There were really no Indian Americans on TV or in the movies at the time, so I didn't see any representation of myself in the world to help me figure that stuff out."

This early awareness about race and ethnicity has influenced Ami's research. Women are exposed to more harmful chemicals in personal care products than men are, but Ami's work has shown that these exposures vary by race and ethnicity too.

"Compared with white women, women of color have higher levels of beauty product–related environmental chemicals in their bodies, independent of socioeconomic status," Ami wrote in a 2017 paper.[1] Ami and her coauthor, Bhavna Shamasunder, pointed out that personal care products used by women of color are often particularly full of toxic chemicals, but that studies on environmental risk factors among women of color typically overlook these exposures, focusing instead on polluting industries or high levels of traffic—which misses a critical part of the picture.

"I think it's useful to think about how systems of oppression like racism, classism, and sexism affect our environmental exposures in an upstream way," Ami told me. "For example, there's a global hierarchy of beauty norms informed by colonialism and slavery that puts white women at the top, so white femininity is the beauty ideal that women are told to strive for."

Women throughout Southeast Asia use skin lightening creams, designed to make them look whiter, that contain dangerous chemicals, and Black and Latina women are more likely than their white counterparts to use hair relaxers that are full of toxic chemicals. Research has shown that in the US, Black women are exposed to more toxic chemicals in personal care products than any other racial demographic.[2] A big part of the discrepancy can be traced to hair products marketed to Black Americans, which are part of a long history of stigmatization of Black hair.

Ami pointed to a 2018 incident in which Andrew Johnson, a Black sixteen-year-old wrestler from New Jersey, was informed by a white referee that his hairstyle didn't conform to the rulebook. He was told that he had to either cut his dreadlocks or forfeit the match, so he begrudgingly agreed to cut his hair. A video of a white adult cutting off a teary-eyed

Black teenager's dreadlocks against his wishes went viral and prompted a national outcry. In the US, natural Black hairstyles, certain types of braids, and dreadlocks have historically been banned by dress codes in many schools, workplaces, and the military, and straight hair is often described as being "more professional" for Black women. It's no surprise that African American consumers purchase nine times more ethnic hair and beauty products than other racial groups.

Cosmetics and personal care products of all kinds contain chemicals like parabens, phthalates, formaldehyde, per- and polyfluoroalkyl substances (PFAS), petroleum by-products, siloxane, and heavy metals. Parabens have been widely used since the 1920s as preservatives to stop the growth of bacteria and mold. Like BPA (bisphenol A), they are also endocrine disruptors. Researchers have found parabens inside the bodies of more than 90% of Americans tested for the chemicals,[3] and at greater concentrations in the bodies of people who regularly use cosmetics. A 2019 study found that teenage girls who wore makeup every day had propylparaben (one of several types of parabens) in their urine at twenty times the levels seen in girls who rarely or never wore makeup.[4] The links between parabens and cancer are still being studied, but research has found that as with other endocrine disruptors, reproductive and breast tissue cells seem to be most heavily affected. Studies have found that parabens can switch on cancer genes and speed up the growth of breast cancer cells.[5] The use of parabens in cosmetics is restricted in the EU, Japan, and several other Southeast Asian countries, but not in the US.

The story is similar with phthalates, commonly used as plasticizers to give flexibility to vinyl products, such as shower curtains and raincoats. They're also used in varnishes and adhesives, in cosmetics like hairspray and nail polish to make them more flexible and less brittle, and in perfume and cologne to strengthen scents and make them last longer. Like parabens, phthalates are endocrine disruptors that can speed up the growth of breast cancer tissue. They can also affect reproduction and brain development and are linked to diabetes. In 2008, the US started restricting phthalates in children's toys. Similar restrictions have been enacted in Canada, Austria, Denmark, Finland, France, Germany, Greece, Italy, Japan, Iceland, Mexico, Norway, and Sweden. The EU and Canada have more restrictions on the use of phthalates in cosmetics than the US. In 2014, Ami was lead author on a paper that looked at national trends in phthalate exposure, finding that after specific types of phthalates were replaced in consumer goods on a widespread basis, Americans' exposures to those chemicals steadily declined over a 10-year period.[6] The study was significant because it showed that regulations can have a significant impact on our everyday exposure to harmful chemicals.

One especially pernicious route of exposure can be traced to items historically overlooked in scientific research: vaginal care products. In 2019 a group of researchers tested eleven brands of menstrual pads for toxic chemicals. They found multiple chemicals linked to increased cancer risk: methylene chloride was found in two brands of pads, toluene in nine brands, and xylene in all eleven brands tested. They also found two types of phthalates in many sanitary pads, sometimes at levels significantly higher than what's commonly found in plastic goods. Alarmingly, the researchers also looked at diapers and made similar findings.[7] Tampon tests have also detected chemicals of concern. The skin in and around the genitals is thin and particularly prone to absorbing chemicals. And beyond tampons and pads, a host of other vaginal products like douches, washes, wipes, and lotions that promise to suppress smell and promote "cleanliness" contain endocrine-disrupting chemicals like phthalates and parabens.[8]

The idea that women's genitals must be treated with harsh chemicals to smell "clean" has both sexist and racist roots. Documents from the era of American slavery claimed Black people smelled different than their white enslavers, especially when it came to vaginal odors. "Vaginal odors are stigmatized for all women as a form of sexism," Ami said, "but for Black women this is a place where racism and sexism collide. These supposed differences in smell were used as a reason to cast Black women as hypersexual, often to justify raping them." As a result, she added, "There's still a lot of attention to bodily smells in all women, but especially in Black communities, which is a lingering adaptation to these forms of stigma and prejudice."

In 2015, Ami coauthored one of the first scientific studies to look at the connection between race and exposure to toxic chemicals in feminine care products. It found that using products like vaginal douches can contribute to phthalate exposure, and that, compared to white women, higher rates of douching among Black women contribute to higher levels of exposure to certain phthalates.[9] A long history of institutional sexism has meant that in the past, such studies were not regarded as important, and it was difficult to get funding for them, Ami said, which is something that she and other researchers like her are finally beginning to change.

———

Berry hadn't seen a gynecologist in a few years, so she scheduled that appointment instead of making one with her primary care doctor to check on the lump in her breast, thinking she'd get it all done at once. When she called in July, they scheduled her for their earliest availability, which wasn't until September.

When she finally made it to the gynecologist in the fall, Berry was nervous. Her doctor felt the lump and said she thought maybe it was just a cyst, but she ordered a mammogram and an ultrasound to be safe. Those procedures took another month to schedule, and Berry's worry grew by the day.

In early October 2020 she had a mammogram and ultrasound, but her doctor wasn't in the room so she had to wait for the results.

The techs who performed those services couldn't tell her anything about what the images revealed, but they were incredibly kind—"like way, way nicer than usual," she said, which kicked her anxiety into high gear.

When she got home from the appointment, Berry Googled pictures of malignant breast tumors. They looked just like the images she had seen on the ultrasound screen earlier in the day.

A few days after her ultrasound, a nurse Berry had never met called and confirmed her fears: she had breast cancer.

The nurse explained that Berry had invasive ductal carcinoma, but couldn't tell her anything more until they did additional testing—she didn't know what stage it was and couldn't offer insights about what treatment might entail. Berry had two acquaintances who'd had different forms of invasive ductal carcinoma. One of them had died in her twenties. The other was in remission, and she and Berry were still friends. "That was a terrible moment," Berry said. "To know I had cancer but not know which type it was. I really started to spiral."

Two days after her diagnosis she went in for a biopsy, which involved long needles and left her breast bruised and sore. The biopsy revealed that she had a different type of cancer than either of her acquaintances. Her cancer did not involve hormones, but she had higher than normal levels of HER2, a growth-promoting protein. These cancers tend to grow and spread faster than other breast cancers, but they respond well to targeted treatment.

Berry's tumor was grade 3, stage 1a, meaning it was the most aggressive and threatening type of tumor on the scale, but that she had caught it very early—news that was simultaneously reassuring and terrifying. It was hard not to think about what could have happened if she had ignored the small lump in her breast for any longer.

The biopsy also revealed that Berry did not have the genetic mutation that signals inherited familial risk for breast cancer. That awareness sent her down an anxiety-fueled path of what-ifs, scouring her past for things that could have contributed to her cancer.

Had smoking on and off during her college years done this? Or being exposed to second-hand smoke from her parents' and siblings' cigarettes as a kid? She had renovated a few houses and painted some murals without wearing the right kind of respiratory protection—had exposure to asbestos, sealants, and paint fumes contributed to her cancer? Berry had seen headlines about Pittsburgh having some of the worst air quality in the country. Had breathing industrial pollution for most of her adult life caused her cancer? Her cat had died of lung cancer a few years prior—could there be something in her apartment, or in her drinking water? After learning that women who haven't had children are at higher risk for breast cancer, she felt vaguely guilty—if she had followed a different path, stayed in her hometown, gotten married, and started having kids in her twenties, like so many of her high school friends had, would she be healthy now?

———

As Ami got older, several things happened that helped her flourish. Her family strengthened their connections with a close-knit community of other Indian immigrants, spending weekends and taking vacations with a group of five or six other Indian families, which gave Ami a welcome reprieve from feeling alienated at school. She described those friends as being "like cousins," and she remains close to many of them today. Around the same time, she started reading books that expanded her thinking. She got interested in Jainism, the religion her parents had been raised with. The core principles of respect and compassion for all forms of life, including the environment, resonated with her, and she became a vegetarian like her mother. The family also moved to Fayetteville, a bigger North Carolina city with a more diverse student body, and Ami made a group of friends she described as "fellow nonconformists" who fostered each other's interests in activism and social justice.

Today, Ami mentors young scientists, encouraging them to incorporate social concepts that are important to them into their scientific research, like she has. "I try to be a bridge between siloed communities like scientists and social justice activists," she said. "I think part of that comes from my experiences as a kid feeling like I was a bridge between the Indian and American communities. Being able to do that kind of cross-talk between different cultures has been really helpful in my professional work."

As a teen, Ami was a voracious reader, and she became increasingly bored with school. "I've always been very curious and unwilling to settle for easy answers to big questions," she said. "I was just very dissatisfied with the education I was getting—it didn't stimulate me."

Itching to do something different, she applied for the North Carolina School of Science and Mathematics, a public residential high school in Durham, North Carolina, with competitive entry requirements. She got in and moved to Durham on her own as a high school junior, finishing high school there. It was a revelation.

"They have a unique approach to how they teach science and math that's very hands-on and innovative," Ami said. "I loved it, and it planted a lot of seeds. It was also really hard. Everyone there had been at the top of their class, so I had to reorient myself in that new distribution."

By the time she enrolled at the University of North Carolina at Chapel Hill, Ami said, she knew she was interested in thinking about power dynamics, that she gravitated toward women's issues, and that she was deeply interested in science. She spent her undergraduate career trying to figure out where those interests intersected. She initially wanted to study biology and neuroscience but was turned off by how those courses were geared toward premed—huge classes aimed at weeding people out—so she switched to a major in public health and a minor in anthropology. When she wasn't busy studying, Ami had an active social life. She had a large group of friends, worked as a late-night DJ at the campus radio station, and went to concerts whenever she could. In her junior year she spent a semester working on ecological and cultural projects in Madagascar, which solidified her hunch that she wanted to work with people rather than in a lab.

For her senior thesis, Ami was mentored by the late Dr. Steven Bennett Wing, considered one of the fathers of environmental justice. She also worked with the late Dr. Frances M. Lynn, who founded UNC Chapel Hill's Center for Public Engagement with Science. While Ami was at UNC in the 1990s, the organization was one of the very few to connect scientists with community members concerned about water quality and exposure to toxic chemicals. Ami's thesis focused on community-based, participatory research, which prioritizes input from communities rather than studying them from an academic distance. She graduated with highest honors, and a few decades later, such research is increasingly common in the field.

Throughout her academic career, Ami rarely encountered overt racism or sexism, but she was aware of how stereotypes limited people's expectations of her. And that made her want to subvert those expectations. "Generally Asian women were expected to be quiet and obedient," she said. "In academia that's often led to us being thought of as proficient technocrats, but not looked at or groomed to be visionary leaders."

After graduating from UNC Chapel Hill, Ami went to India for a year to connect with her extended family and her cultural roots and to volunteer with social and environmental justice organizations. Her early experiences abroad were influential in her career. "My time in India and Madagascar helped me see clearly how intertwined the health of the community and the health of the environment are," she said. "That connection was more explicit in those contexts, and that led me to environmental health because I realized I wanted to work at the intersection of people and the environment."

While living in India, Ami applied for graduate public health programs in the US. She got into several programs, including one at Harvard's T.H. Chan School of Public Health. Thanks to her undergraduate thesis, Ami was resolute about conducting public health research *with* vulnerable communities rather than *on* them. During her interview with a faculty member at Harvard, she announced that she would only attend if she was permitted to do this type of work. "Looking back now, I'm not sure how I would feel if a prospective student took a similar approach with me," Ami said, laughing. "But luckily he said yes."

Before looking at beauty products and cosmetics, Ami did work related to toxic chemicals in various other parts of people's lives, including homes, neighborhoods, and workplaces. Along with a group of other researchers, Ami embarked on a two-year project that investigated environmental conditions in Boston public housing, tried to improve them, and then measured the results of their interventions on children and caregivers. They also trained residents in how to be community health advocates. Many of the changes—things like cutting down on the use of toxic chemicals for pest control and introducing mite-resistant mattresses—led to measurable improvements in residents' health and quality of life, including fewer asthma attacks and allergy symptoms among children. Ami's first publication, which came out of that project, was the first study to document the impacts of ventilation on indoor air pollution in public housing.

Environmental health is a relatively small field, and many of the scientists doing this work know one another. As a doctoral student, Ami worked with Laurel Schaider investigating heavy metal exposures near a former mining site in rural Oklahoma. While Laurel focused on tracking compounds like lead, zinc, cadmium, arsenic, and manganese in plants and soil near the Superfund site, Ami tested houses in the area.

She found that particles were drifting from the big hills of mining waste into people's homes, where they settled in house dust and, in some cases, contributed to higher levels of exposure in pregnant mothers and children. Her research was among the first to suggest that the millions of Americans who live close to Superfund sites are likely being exposed to toxic chemicals that migrate from those sites into their homes.

"I basically became the lead of my own study, collecting my own environmental samples, doing my own lab work, my own statistics, enlisting help from all of these grandmas in Oklahoma to help me go into the homes of local people," Ami said. "I went back and forth to Oklahoma for two years. It was hard work, but also a very comprehensive training experience."

After earning her PhD in 2007, Ami did a postdoctoral fellowship at the Silent Spring Institute (she later referred Laurel), where she studied endocrine-disrupting chemicals. She specifically looked at polybrominated diphenyl ethers (PBDEs), endocrine disruptors that have been linked to thyroid disorders, breast cancer, and neurodevelopmental harm in babies. PBDEs are used as flame retardants in everything from plastic and electronics to fabric and furniture. Some of the most harmful PBDEs have been restricted in the US since 2004, but the phase-out has been slow and some are still in use. After finding levels of PBDEs in house dust and the bodies of California residents that were nearly twofold higher than national averages, Ami and her colleagues speculated that this could be due to a unique 1975 state law that required the use of flame retardants like PBDEs in furniture sold in the state.[10] "It was a big finding that got a lot of press and there were immediate policy implications," Ami said.

She soon started getting phone calls from politicians who wanted to change the law. Interest grew further after she published a study showing that low-income pregnant people in California had elevated levels of flame retardants in their bodies compared to levels elsewhere.[11] Ami testified about her research several times at the California state legislature, which eventually updated the law to ensure that furniture was resistant to fire without being chock-full of toxic chemicals.

The experience helped Ami realize that her work could prompt policy changes, which were much more effective at protecting people from toxic exposures than anything individuals could do on their own. She was hooked.

She soon learned that other places are ahead of the US when it comes to saving lives and public health spending through the regulation of toxics in personal care products. The EU has banned or restricted more than 2,400 chemicals in cosmetics, while the US has banned or restricted just 9 chemicals. According to a 2019 analysis by the Environmental Working Group, more than forty nations have enacted more stringent regulations on chemicals in cosmetics than the US.[12]

American lawmakers face tremendous pressure from chemical manufacturers to keep regulations lenient. The history of asbestos regulation is a prime example. The EPA attempted to ban asbestos—which is highly carcinogenic—in 1989, but the ban was overturned by a federal court in 1991 following backlash from manufacturers. Asbestos is a naturally occurring mineral that's often found in the same places where talc is mined. Talc is a common ingredient in cosmetics and personal care products, and asbestos contamination in talc is not uncommon. Today asbestos can still sometimes be found in makeup sold in the United States, decades after being banned in the EU and Canada. In 2019, for example, the US Food and Drug Administration (FDA) warned consumers that it had found asbestos in eyeshadow, contour palettes, and compact powder in Claire's cosmetics, which are marketed and sold to tween and teen girls, but the agency had no legal means of pulling the products from the shelves. The company pulled the products voluntarily, but initially refused the FDA's request to issue a recall, which requires additional steps, such as warning consumers.

The company only voluntarily recalled the products after getting some bad press.

As part of her push to advance better regulations in the US, Ami has testified on Capitol Hill about the dangers of PFAS in cosmetics, the need for scientific research on toxic chemicals in beauty products, and the reasons current regulations are failing to protect Americans. "In the US we're just not very good at developing health-related policies focused on prevention," Ami said, noting that the Affordable Care Act, passed in 2010, was the first piece of legislation to include a national strategy on disease prevention. "Realistically, we're going to have to address toxic chemicals in a piecemeal fashion rather than all at once."

In 2019, Ami testified at an informational hearing that ultimately led California to ban twenty-four carcinogenic and toxic chemicals, including mercury, formaldehyde, and some types of phthalates, parabens, and PFAS, from cosmetics sold in the state. The law was passed in 2020 and will go into effect in 2025. Ami expects it to have a sizable impact. "It's hard for companies to have different manufacturing processes for different states," Ami said, "so there's a high likelihood that these restrictions will also become de facto for the rest of the country. It's a big victory."

———

During the pandemic, rates of anxiety and depression in the US spiked, from 11% of adults affected to about 41%.[13] Berry was among those affected. She had struggled with depression in the year leading up to March 2020, and it had gotten worse as the pandemic dragged on. "I wasn't suicidal," she said, "but I had what my therapist described as 'passive thoughts of death.' Like it wouldn't really matter if I wasn't here anymore." Her breast cancer diagnosis immediately changed her perspective. "I knew right away that I didn't want to die," she said. "It kind of snapped me out of it." Still, the weeks after her diagnosis were a blur of fear, anxiety, and grief. They were also physically painful.

Within a few days of her diagnosis, Berry's doctor explained that che-motherapy can make it difficult to conceive a baby, and instructed Berry to decide as quickly as possible whether she wanted to freeze her eggs. If she did, her cancer treatment would have to be deferred until after her eggs had been harvested, and it was critical to start treatment as soon as possible.

It was a big question to grapple with—she had always been ambiv-alent about having kids—but within a day, Berry decided she at least wanted to have the option of having a baby down the road. She imme-diately started fertility treatments, which involved taking injections at home three times a day. But Berry found herself unable to plunge the needle into her own abdomen or thigh as she had been instructed to do. She froze with fear and dread every time she tried. She asked her boyfriend to help, but he was too squeamish and afraid of hurting her. In the end, a neighbor who was a nurse and a friend who was a doctor agreed to help out. Berry either went to their homes or they came to hers three times a day to give her the shots, which prompted intense hormonal swings that made her moods volatile.

Immediately after her diagnosis, Berry had started a blog to docu-ment her journey and share health updates with family and friends. In her entries during this time she expressed gratitude for the help admin-istering the shots, but also wrote that they made her feel "like a bloated pin cushion."

After about two weeks of fertility treatments, Berry underwent sur-gery to harvest her eggs. While she was sedated, Doctors pushed a needle through her vaginal wall and into one of her ovaries to remove thirty-two eggs, twenty-two of which were deemed viable and were fro-zen for future use.

Berry's boyfriend picked her up afterward. On the ride home, achy and drowsy, she vomited into a bag in the passenger seat, an aftereffect of the anesthesia. Just three weeks ago, even though a cancerous tumor was growing in her breast, she had felt like she was in perfect health. Now, she hadn't even begun cancer treatment, and already her body felt hobbled and depleted.

Two days after having her eggs harvested, Berry had a small plastic port for chemo medications surgically implanted in her chest. She was wary about undergoing anesthesia again so soon after the egg harvesting procedure so she opted for local numbing instead. She didn't feel pain, but the procedure was uncomfortable and unsettling. "You can feel tugging and jamming and hear all the noises the surgery is making," she said. "I actually really wished I'd been asleep for it."

Berry's chemotherapy port was placed on the left side of her chest. Her right breast, the one with the tumor, still hurt from the biopsy, so she was sore on both sides and had trouble sleeping after the port was placed. During that time of little sleep, her anxiety returned. She channeled it into things that she could control: she quit drinking after reading a study linking alcohol consumption with higher breast cancer risk, started eating healthier, and bought natural personal care products that were free of parabens and phthalates. She had already put in a water filter after her cat died, and she had done a radon test when she moved in, but she did another one just to be safe (it came back fine), and she had the ductwork in her home professionally cleaned. Some of these things made her feel empowered, but it was a fine line to walk, feeling like her choices controlled her fate. If she missed a day of exercise or slipped up and ate some junk food or had a beer, she would feel overwhelmingly guilty, afraid that she might be making her cancer worse.

This kind of confusion and self-blame is common among people who are diagnosed with nongenetic cancers. It's hard not to wonder, if they had made different choices—about everything from where to live and what to eat and drink to which shampoo to buy and what type of livelihood to pursue—would they still have gotten cancer? If they started making different choices now, could they increase their odds of survival, or their odds of staying cancer-free in years to come?

Berry will never know for certain what factors contributed to her cancer. It's possible that if she had never smoked cigarettes or drunk alcohol, her breast cancer would never have developed. It's also possible that she never would have gotten cancer if she had grown up in a place with fewer carcinogens in the air, or if she had never eaten food treated with pesticides, or if she had lived in a part of the world where phthalates and parabens were banned in lipstick and lotion. "The uncertainty is really scary," Berry said. "It just makes you feel kind of powerless."

———————

Ami sees four main ways to reduce people's overall exposure to cancer-causing chemicals: educating consumers, shifting marketplace demands, changing policies, and challenging social norms related to beauty.

"The most equitable solutions will always involve policy changes that protect everyone," she said. "But we understand that that's the long game and people want to feel like they have control over this situation sooner, so we also offer individual actions that can help."

For instance, you can avoid personal care products that contain phthalates, parabens, formaldehyde, and heavy metals, like Ami does, to minimize your exposure. You can go further and advance market changes that will help protect others at the same time by telling companies using those ingredients that you've stopped using their products and why, and urging them to switch to safer alternatives.

"Often luxury brands tend to move first on these things because people with more money and education tend to be more vocal about what's in their consumer products," Ami said, "but there are also other ways people are using market-based strategies to help create change from the bottom up."

One example is BLK + GRN, an online marketplace featuring toxic-free skin, hair, bath, menstrual, and self-care products designed by small, Black-owned companies and independent Black artisans. The marketplace was founded by Kristian Edwards, who holds a doctorate in public health from Johns Hopkins University. After learning that Black women are disproportionately exposed to toxic chemicals in personal care products, Edwards started ridding her own cosmetics collection of those chemicals around the same time that she was reading a book called *Our Black Year*, about a Black woman's experience trying to purchase from only Black-owned businesses for a year. Her desire to make healthy products more accessible and support Black-owned businesses spurred the creation of BLK + GRN. She developed a list of specific ingredients that are banned from the marketplace, listing only vendors that make products free of those ingredients on the site.

This kind of market pressure can help and can be empowering, Ami said, but it's not a complete solution. "You really can't completely shop your way out of the problem of toxics," she said. "They're everywhere, so it's virtually impossible to buy toxic-free versions of everything you need. It's also cost-prohibitive, and it becomes really mentally draining, so only people with disposable income and the intellectual bandwidth to deal with this kind of stuff even have the ability to try."

For herself, Ami tries to buy organic and less-processed food and picks a few things to avoid—namely flame retardants and toxics in cosmetics and personal care products—but is pragmatic about the rest. "I try to minimize risk, but within reason," she said. "I'm not going to pay three times the cost for something to try to go from 10 parts per billion of exposure to zero parts per billion. The relative cost versus the relative gain isn't always worth it. And knowing that these options aren't available to most people makes me not want to be too elitist about it."

"Most studies on this have found that individual actions can't reduce exposures to zero anyway," she added. "Looking at phthalates, for example, it's hard to even figure out the dominant source of exposure in someone's life—is it food, beauty products, or air pollution? The thing that's going to help people the most for the longest term is policies that regulate these chemicals."

She pointed to the Clean Air Act as an example of effective regulation. In terms of lives and money saved, it's the most powerful public health law ever enacted in the United States. In 2020 alone, more than 230,000 early deaths were prevented by the Clean Air Act, and since its passage, it has prevented millions of deaths and saved trillions of healthcare dollars.[14]

In order to advance meaningful science that will lead to sweeping regulatory changes, Ami said, researchers will need to incorporate intersectionality—the idea that different systems of oppression, like racism and sexism, overlap to amplify harm—into their work. Intersectionality isn't new in the social sciences, but Ami was among the first researchers to suggest incorporating the concept into environmental health work. In 2020, she published a paper with coauthor Brianna N. VanNoy urging the environmental health research community to consider the interconnected effects of sexism and racism in public health studies, using their recent investigation on uterine fibroids in Black women as an example.[15]

Uterine fibroids, noncancerous growths in the uterus, are the most common tumor in women: 70% of white women and more than 80% of Black women develop fibroids, usually before reaching the age of fifty. Sometimes they're symptomless, but up to 30% of women with fibroids develop symptoms like pelvic pain, heavy bleeding, pregnancy complications, and infertility. The tumors can be removed but are prone to growing back, so a hysterectomy is the only permanent treatment. Most are benign, but it's possible for fibroids to be cancerous. Black women experience a higher risk of fibroids, tend to develop them at an earlier age, and tend to have more severe symptoms, but scientists and doctors are still mostly in the dark about why—and not for lack of trying to figure it out.

Historically, most scientists have tried to identify molecular and genetic causes of fibroids, focusing on biological differences that might drive these disparities. Some of those studies have detected biological differences in fibroids from different racial groups, but a 2018 study concluded that these differences don't actually explain why Black women are more likely to develop them or experience worse symptoms.[16] Several studies have linked the use of hair relaxers to higher rates of fibroids, so Ami and her colleagues are investigating whether endocrine-disrupting chemicals could be a contributing factor. In preliminary studies, they found that among women undergoing surgery for symptomatic fibroids, phthalate exposures were higher in Black women compared to white women, and exposure to certain phthalates was associated with more severe cases of the disease. These studies were small so more research is needed, but they suggest that phthalates may contribute to uterine fibroids and offer insights into racial disparities.

If they weren't aware of how racism and sexism affect the products women use, Ami said, they would never have undertaken such an investigation or made such a breakthrough. Her paper on intersectionality urged others to consider similar factors in their own work. "Clinicians and researchers must be trained to not only address their own biases," the paper concluded, "but also recognize and address the historical and social-structural context within which Black women seek clinical care."[17]

In the months after the paper was published, Ami was asked to speak about it at seminars across the country. This kind of work isn't very common, Ami said, so there's a lot of interest in it. She hopes to leverage that interest to push her fellow researchers to engage in antiracism work—meaning not just talking about racial inequity, but actively working to end lingering racial stigma and dismantle racist systems.

As part of her focus on equity, Ami has partnered with environmental justice organizations in New York and Los Angeles for projects aimed at exploring beauty care norms among people of different races and ethnicities. Ami is a co-investigator on the Taking Stock study, led by Dr. Bhavna Shamasunder at Occidental College and Janette Robinson Flint, executive director of Black Women for Wellness, a nonprofit health advocacy organization. The study surveyed several hundred women throughout California about their use of cosmetics and personal care products. It found that women reported using a median of eight products daily, with some women using up to thirty products daily, and that Hispanic and Asian women used more cosmetics, while Black women used more hair and menstrual/intimate care products than other women. The project is ongoing, and the researchers are also investigating what kinds of products are available in retail stores where the women live and what drives their purchasing choices, hoping to gain more insights about the ways racism, classism, and sexism affect who buys which products. For a similar project in New York, Ami is working with the environmental justice nonprofit WE ACT for Environmental Justice to survey femme-identifying people in the South Bronx and Northern Manhattan about the types of cosmetics they use and their attitudes about those products. The ultimate goal, Ami said, is to develop local and state-level policies and community-informed education that tackle the environmental injustice of beauty.

Ami wants to keep expanding our understanding of environmental health. Next, she hopes researchers will start considering people with disabilities and members of the LGBTQ+ community when thinking about neglected and vulnerable populations. She also wants to see more research on the combined impacts of harmful exposures from people's workplaces, neighborhoods, and use of personal care products, with an eye toward the ways that systemic social problems contribute to our rates of exposure. For example, it's important to recognize that low-income people are more likely to hold janitorial or landscaping jobs that expose them to toxic cleaning products or pesticides and that polluting industrial sites are more likely to be sited in low-income neighborhoods. But it's equally important to recognize that neighborhoods where people of color live also tend to experience more air pollution than predominantly white neighborhoods, regardless of income levels.[18] In part, this is due to historical practices like redlining, which limited where people of color could buy or rent property.[19] Those same communities of color may also use a higher volume of personal care products that contain harmful chemicals as a result of racist beauty norms, and it's important to understand how these disproportionate exposures accumulate and what forces are driving them.

"Now that I'm at a point in my career where I have more choice about what I work on, I think a lot more explicitly about how we can push the field of environmental health to think differently about these issues and connect the science to real solutions," Ami said. "Trying to get the field to be antiracist is an aspirational goal, but it's very exciting to be doing science aimed at driving social change."

CHAPTER 3

Nse: Safer Little Ones through Politics

Berry's hometown, Oil City, sits in the foothills of the Appalachian Mountains where the Allegheny River and Oil Creek converge in Venango County, about 70 miles north of Pittsburgh. The area was once a Seneca Indian village named Onenge, meaning otter or mink, which white colonizers misheard as Venango. The chief of the Wolf Clan of the Seneca nation, Cornplanter, was given back a tract of land that's now part of Oil City (along with larger tracts of land elsewhere) as payment for helping with negotiations after the American Revolutionary War, but the county later attempted to retake the land from him for nonpayment of taxes. Today, monuments to Cornplanter can be seen around town in Oil City, which was named for its role in the birth of the American oil industry.

On August 27, 1859, one of the world's first oil wells was drilled in Titusville, about 16 miles north of Oil City. Oil City sits alongside a bend in the Allegheny River. The bend slows the currents down, making it an ideal place for barges and boats to land, so it soon became a shipping hub for the burgeoning industry. In the town's heyday, riverboats transported millions of barrels of crude oil from Oil City to Pittsburgh.

Pennzoil, Quaker State, and Wolf's Head motor oil companies were all founded in Oil City.

The local population grew steadily until it peaked at around 22,075 in 1930. It then began a slow but steady decline as the oil industry fell into its ongoing cycle of booms and busts. By the 1990s, Pennzoil, Quaker State, and Wolf's Head had all relocated their headquarters elsewhere. Today, only around 9,500 people live in Oil City, including Berry's family.

———

Nsedu Obot Witherspoon (or Nse, pronounced EN-say, for short) has spent twenty-two years working to keep harmful chemicals away from the people most vulnerable to them: children. She's the executive director of the Children's Environmental Health Network (CEHN), a non-profit based in Washington, DC, that was started in the 1990s. She has also served as an adviser for the CDC, the EPA, and the National Institutes of Health (NIH), and Nse is a co-leader of the science and health arm of the Cancer Free Economy Network.

Relative to their size, children breathe more air, drink more water, and eat more food than adults, and their systems are still developing. This means that their bodies are less efficient at filtering out toxic chemicals than adult bodies are. It also means that delicate, hormone-sensitive processes that are under way in kids can be interrupted by endocrine-disrupting chemicals, throwing a wrench in development and raising cancer risk. Emerging science also indicates that parents' exposures to chemicals before and during pregnancy can increase the risk of childhood cancer,[1] and that kids' earliest exposures increase their likelihood of developing cancer later in life.[2] Despite the science, most American schools and daycares are filled with harmful chemicals that can make their way into kids' bodies from old materials in school buildings (e.g., polychlorinated biphenyls or PCBs, formaldehyde, and lead-based paint), pest-control and cleaning products, flame retardants in carpeting and furniture, and PVC (polyvinyl chloride) plastic used to make chairs, toys, and utensils.

Nse has four children of her own. She gave me a virtual tour of the home she shares with her husband and kids in the DC area in 2021. With kids spanning the ages of five through eighteen there was rarely a quiet moment. The kids teased each other, talked over one another, and guffawed over inside jokes. Over the years they'd collected a bit of something for everyone: a ping-pong table, a piano, video game consoles, Nerf guns, soccer balls, bins of dolls and blocks and puzzles and games, a punching bag, crystal and mineral collections, an air hockey table, a vast quantity of Pokémon cards, and boxes of hand-me-downs. They even have a swimming pool out back.

Nse, who has golden-brown skin and a wide, contagious smile, said she's always had an affinity for children, but having kids of her own amplified her protective instincts. "Inevitably you're mostly focusing on yourself until you start caring for other little souls," she said. "Becoming a mom made me feel even more strongly how important it is that we protect these vulnerable little people who aren't yet able to stand up and protect themselves."

Through CEHN, in 2004 Nse helped launch one of the first programs in the US aimed at reducing chemicals in early childcare facilities like daycares and preschools. A year later the Oregon Environmental Health Council launched a similar state initiative, and in 2010, the groups merged the two to create the national Eco-Healthy Child Care Program. At the outset of these projects, Nse said, there were already several programs aimed at preventing harmful chemical exposures in K–12 schools, including the EPA's healthy school environments toolkit and green school certification programs run by state agencies and various nonprofits—but there were none for children younger than elementary school age.

Initially, some of CEHN's board members were skeptical about whether such a program could be effective given the scale of the problem. They started small, launching pilot programs in a few states that offered free trainings for childcare providers on how to reduce chemical use, and why it was important. Nse served as a trainer during the program's early days, when demand quickly grew through word of mouth. "It was really powerful to be in places from California to rural Georgia, watching people have the same kinds of reactions to this information we were providing," she said. "There's a whole emotional cycle that happens when people find out they've been exposed to harmful chemicals their whole lives without their permission. First there's anger, then guilt, then usually more anger. There's this false thinking that as Americans, we're more protected than we actually are."

Today, Nse said, the program remains the only one in the country focused on preventing toxic exposures at early childcare facilities. It's available in most states and has trained more than 2,200 childcare professionals. It has also won several awards, including a 2019 Clean Air Excellence Award from the EPA.

The program certifies facilities that comply with at least thirty of thirty-five free or low-cost practices. The checklist, available online, is detailed and comprehensive. It includes things like using nontoxic or less-toxic pest control and cleaning methods, ensuring proper ventilation, barring idling in parking areas, testing drinking water and soil for lead, choosing the least-toxic toys and art materials, and switching to safer materials for furniture, paint, and carpeting when renovating, among many others. Trainers in the program also review the health threats posed by chemical exposure: learning, developmental, and neurological disabilities; reproductive harm; asthma; and cancer. "We're not at all into scare tactics," Nse said. "But we believe people have the right to know these things so they can take steps to protect themselves and the children in their care. First we empower them to do that, then we try to help them take their angry energy and pivot it toward pushing for solutions on a large scale."

As part of the program, staffers may make surprise visits to participating facilities to ensure that they're actually implementing the changes, and to offer help for those that are struggling. The program is voluntary—childcare centers sign up, then they can advertise that they've received an endorsement through the program—but Nse hopes that eventually it will become part of state licensing and accreditation programs. "What happens now is that there are sometimes state recommendations, but nothing is required," she said. "Ideally, avoiding toxic chemicals in these facilities should be a health mandate—facilities would either meet legal safety requirements or not, and there would be tools and programs in place to help the ones that can't meet the mark on their own."

Groups as far away as South Africa and Australia have reached out to Nse and her colleagues for help launching their own similar programs, and the Children's Environmental Health Network has shared guidance and technical assistance. "I'm really proud of this program," Nse said. "It's the nature of prevention that we'll never get to know the millions of kids we've helped. But knowing that we did is enough."

Another aspect of Nse's work is pushing for stronger pesticide regulations at both the state and national level. Products aimed at killing rodents and insects in schools and childcare centers often contain chemicals like arsenic, ethylene oxide, and lindane, and numerous studies have shown that people who work in pest control[3] have a higher cancer risk, as do their children.[4]

In recent years, Nse has been advocating for a bill that could radically reduce American kids' exposure to pesticides. The bill, originally introduced as the Protect America's Children from Toxic Pesticides Act of 2020 (PACTPA), would represent the first comprehensive update to the main law governing pesticide use in the United States since 1996, when the Federal Insecticide, Fungicide, and Rodenticide Act, or FIFRA, was updated. PACTPA would ban three of the most dangerous categories of pesticides: organophosphate insecticides, which are linked to neurodevelopmental damage, lymphoma, and leukemia in children; neonicotinoid insecticides, linked to birth defects, endocrine disruption, and breast cancer (as well as threatening critical pollinator insects around the world); and the herbicide paraquat, linked to Parkinson's disease and non-Hodgkin's lymphoma. These pesticides have already been banned in many other parts of the world.

PACTPA would also close loopholes that allow new pesticides to enter the market without full health and safety reviews, allow citizens to petition the EPA to identify dangerous pesticides so they don't indefinitely stay on the market, suspend the use of pesticides deemed unsafe by the EU or Canada until they've been thoroughly reviewed by the EPA, and allow local communities to regulate pesticide use without being overridden by more lax state or federal laws—something that currently happens routinely. The senator who originally introduced the bill retired before it could be voted on, but Senator Cory Booker replaced him as the bill's lead sponsor. In the summer of 2022, Nse and her colleagues were looking for a House leader to help push for the bill's passage (which might happen under a new name).

"The federal government doesn't do things overnight, but even so we're lagging way behind the rest of the world on this," Nse said. "Many times with toxic chemicals, the science is behind what we already know from common sense and personal experience. But we have twenty-five years of science on harms to children's health from pesticides at this point. It's frustrating."

While they continue pushing for federal regulations, Nse is also working for change at the state level. Right before COVID-19 hit she joined other health advocates in Maryland to testify about the risks posed by the pesticide chlorpyrifos for lawmakers considering a statewide ban. Chlorpyrifos is used in farming and the maintenance of golf courses, and exposure is linked to brain damage in children and fetuses, increased risk of autism, low birth weights, endocrine disruption, and increased risk of lung and prostate cancer. Until 2001, chlorpyrifos was also used in household pest control products, but it was banned from those products after it was discovered that kids could easily be poisoned by it. When Nse traveled to Annapolis, Maryland, in 2019, thirty-five other countries had already banned the pesticide, and some states, including Hawaii and California, had followed suit.

"We had more than a dozen pregnant mothers, young people, and parents with brain-altered children give incredibly moving testimony about the harm this pesticide had caused them and their families, and the fear they were living with while it was still in use," Nse said.

"Then six or seven people from the industry got up and talked about things like how chlorpyrifos is needed to keep golf courses nice. This had been happening in state legislatures all over the country, industry reps testifying that this specific pesticide is a benefit to society."

Lawmakers were moved by the testimony of the families, so the bill with the ban passed easily. Nse, her colleagues, and the families were still celebrating when Maryland's Republican governor vetoed it, basically sending them back to square one. "There are moments where it's tempting to throw up your hands, but instead we have to dust ourselves off and start again," Nse said, noting that it's emotionally difficult for affected parents and families to get up and testify during public hearings, and it's asking a lot of them to have to do it repeatedly.

"We usually do eventually see success," she said, "but it's still frustrating to think of how many lives could have been saved and improved if it passed the first time we tried."

———

Present-day Oil City looks like many former American boomtowns that have slowly decayed through decades of bust. An anachronistically quaint main street is flanked with regal, once-bustling buildings that now sit empty, their windows boarded up, some plastered with signs offering redevelopment opportunities.

As Berry and I looked through the windows of the former National Bank on a sunny day in December 2021, she pointed out the ornamental marble floors, visible in patches through a thick layer of dust, and the crumbling remains of an intricately painted ceiling. Berry was still undergoing radiation treatments, but chemo was well behind her and her hair had grown into a cute pixie cut, her soft curls peeking out beneath the ochre beanie she wore. Town officials had been trying to rehab the National Bank building and turn it into a microbrewery for years, she said, but the plans kept getting delayed. Across the street, she pointed to where an abandoned department store had been demolished a few years earlier and replaced with a small park. There were cheery Christmas bows on the benches and lampposts and a large blue receptacle labeled "Santa's Mailbox."

Just up the street is the National Transit Building, a four-story stone and brick structure with elegant archways over its doors and windows. It was once the hub of John D. Rockefeller's Standard Oil Company, where the world's oil prices were set. In recent years, the building has been repurposed into discount studio spaces through the town's Artist Relocation Program, which entices artists to move to Oil City through low-interest home loans and other incentives. Thanks in part to the program's success, murals are painted on the walls of numerous downtown buildings, including one painted by Berry. Berry went to Allegheny College, about forty-five minutes from Oil City, where she majored in international studies and fine art. Her studies took her to Germany, Paris, New Zealand, and Ukraine for an unusual combination of business, drawing, and painting courses. She paints on canvas sometimes, but most of her paid work has been in the form of large-scale murals throughout western Pennsylvania.

Berry's Oil City mural is a block-long painting on the side of Billy's, a red brick building that houses a bar and restaurant on the first floor and apartments above. It depicts red stage curtains parted to reveal a nearly virgin Oil City—green, rolling hills cut through by a pristine blue river, an accurate rendering of St. Joseph's, a historic red church on top of a hill, and glowing yellow street lamps and fireflies. On the right side of the mural are two Americana folk musicians born of Berry's imagination: a seated man with a mustache and glasses in a blue suit with a red tie beside a standing woman in a blue skirt, red shirt, and white hat, her curls blowing in the wind, both of them playing acoustic guitars. On the left side of the mural there's a trolley car that Berry has filled with people she loves. Her dad as a young man, looking cool in jeans and a suit jacket, leans against the door. Her grandma, wearing a straw hat and a blue dress with a wide lace collar, looks out from one of the windows, and her brother, wearing a tux, sits in the rear of the car. Billy, the owner of the building, is an old family friend, so Berry also painted Billy's dad in the role of a smiling conductor.

Out of place in the otherwise idyllic mural are the oil derricks— twelve of them peppering the rolling hills, each with intricate crisscrossing shadows painted against the bright green grass, one of them with innocuously cartoonish drops of oil spurting out the top.

———

When Nse gave me a virtual tour of the CEHN office in February 2021, she showed me the view out the window. The US Supreme Court and Capitol buildings were visible across the street, surrounded by heavy metal gates with barbed wire around the top—a new feature following the January 6th riot at the Capitol, when a group of Donald Trump supporters broke in and injured more than 140 police officers. Nse was going into the office only once a week due to COVID-19, usually as the only person there, and the sight of the barbed wire and the National

Guard and Capitol Police officers patrolling the block still made her feel uneasy. The office had recently been renovated so most of the staff's things were stacked in boxes, but the walls were painted a warm yellow, and there were plants on the windowsills—cheery touches set against the guns, gates, and barbed wire outside. The Children's Environmental Health Network has just six full-time staff members, plus a rotating group of interns, fellows, and consultants, all of whom were working from home during the pandemic.

When COVID-19 arrived, Nse and her colleagues were asked by the CDC to work alongside other public health groups to create guidelines about ventilation, handwashing, and cleaning versus sanitization versus disinfection. "Cleaning" refers to physically removing dirt and germs with soap and water, which can help minimize the spread of the virus, while "disinfection" refers to using chemicals that kill most germs (most often called for in situations where blood or other bodily fluids are present), and "sanitization" means using chemicals to lower the number of germs to a level considered safe by public health standards.

"We've tried to provide resources for parents to let them know that being scared about COVID doesn't mean the best option is to spray everything with bleach, which can be harmful to kids, and that there are safer options that will still provide protection against the virus," Nse said, explaining that bleach can worsen or initiate asthma and may cause a buildup of chloroform in the air, which increases cancer risk. "A lot of our regular messaging has been amplified by the pandemic, but we've tried to make sure we're not being tone-deaf and are providing a helpful service that addresses parents' most pressing fears and worries."

After I'd seen her actual office, Nse's kids Ajani, Adiaha, Ahian, and Aryanna also gave me a tour of her home office via Zoom. At the time, Nse's oldest son, Ajani, was looking forward to his high school graduation, and her youngest daughter, Aryanna, was still in preschool. The three older kids were doing school remotely due to the pandemic, and both Nse and her husband, Sikarl, were working from home, so as part of the tour they showed me the corners of the house where each liked to hide out when they needed to focus. Nse's work was spread out all over the dining room table, while the desk allocated to her in the office sat empty, which the kids teased her about. "I don't know why I work better at the dining room table, but I do!" Nse said, laughing along with them.

While it might seem obvious that having a career devoted to protecting children would give her some bonus skills as a mom, Nse said motherhood has also made her better at her job. "Sometimes the best thing to do is take off my work hat and put on my mom hat," she said. "Sadly, this work brings me into contact with so many children who are sick with illnesses that could have been prevented. I'm always aware that just as easily could have been one of my children. My ability to connect with people on that level, just parent to parent as people who want safe, healthy lives for our kids, is a big part of how I'm able to be effective."

When it comes to avoiding toxics in her own home and on behalf of her own kids, Nse said that even she still feels overwhelmed sometimes. "I wish everyone had access to the same resources I did, and even having those it's still hard," she said. "Can we afford to have our entire grocery bill be organic? No. If there are things they eat multiple times a day or that I know are particularly likely to hang onto pesticides, I prioritize those things. I spend a lot of time reading labels but I don't make myself crazy. It's really difficult trying to do everything right—it highlights the need to make this country safer for everyone so you don't have to work so hard to protect your kids."

Like her desire to protect her own children, Nse's family history also drives her in her work. She was born in 1975 in Hammond, Indiana, about 25 miles southeast of Chicago where both of her parents worked. Hammond was more affordable, which Nse realized later was because it was highly industrialized and therefore more polluted.

Nse's father, Otu Asuquo Obot, came to the US in the early 1970s when the Biafran Civil War made Nigeria unsafe. He enrolled in college in Indiana, where he met his wife, Carol Ruth Obot, whose grandparents had immigrated to the US from Germany. When Nse was about eight months old, Otu was homesick and the civil war had ended, so the family moved to Nigeria, where both of her parents took teaching jobs. They stayed for about a year, but when Nse's mother got pregnant again in 1977 the couple wanted to return to Carol's childhood home in Buffalo, New York, so her parents could help support their growing family.

Carol moved first and Otu planned to follow after wrapping up his teaching year. But their plans were interrupted when US immigration officials accused the couple of faking their marriage for a green card. While officials deliberated on their case, Otu was stuck in Nigeria. Carol gave birth to Nse's younger sister, Ekaete, while he was still there. Carol begged her relatives to make phone calls and write letters to their elected officials pleading for help getting Otu back to the US. They did, but nothing seemed to help. Days apart stretched into weeks and then months as Ekaete grew without ever having seen her father's face. Nse was getting bigger too, and Carol worried about the state of her family. "I don't remember much from that time," Nse said, "but I know it was really hard on both of my parents."

Nse was nearly three years old and her sister was nearly six months old by the time Otu was allowed to return to his family. She was too young to remember her parents' reunion, but she imagines it as tearful, joyful relief. When she thinks about it now, she's outraged on her parents' behalf. They've recounted how they were subjected to separate interrogations intended to catch them in a lie. "They were asked things like which side of the bed each person slept on and what brand of toothpaste they used," Nse said. "It was so belittling and offensive. Do people who fake a marriage for immigration purposes really go so far as to have two kids together?"

The family began the process of rebuilding a life in the US. A few years after Otu's return the couple had a son, Nse's little brother, who shares a name with his father. Nse remembers long, warm days spent riding her bike around her suburban neighborhood outside Buffalo, where she felt safe and independent. She took free tennis lessons in the summer, and she and her sister walked the few blocks to their elementary school every day during the school year. They spent a lot of time with their maternal grandparents, who lived just a few minutes away.

"My mom really did her best trying to raise mixed-race children and embrace the culture," Nse said, adding that after what her father had gone through to move to the US, her mother also did her best to make him feel at home. "She often wore Nigerian clothing and learned to cook Nigerian food, things she still does to this day."

Otu was the oldest of six, and he eventually convinced all but one of his siblings to move to the US. Over the years Nse got to meet new cousins as the extended family moved to the states, many of them settling nearby in western New York. The family was heavily involved in the community, and Nse was outgoing and popular—she sang in her school choir and played a different sport every season—but she was also acutely aware of the ways in which her family was viewed as different.

"I remember sleepovers when I was a teenager where girls would ask a lot of questions about my hair and want to touch it," she recalled. "I knew they weren't being spiteful. They were just inquisitive, but looking back now I think I felt a bit like a guinea pig. I knew my mom and my dad looked different and that I had different skin and hair than most of my classmates."

Her upper-middle-class public school was predominantly white, but the University of Buffalo was home to many international students, and Nse remembers seeing that racial diversity outside her school and feeling appreciative for that sense of community.

"They protected us from it, but when I think about my parents in the seventies in the Midwest, I think it must have been hard for them," she said. "They weren't just Black and white, but Nigerian and American. That kind of thing wasn't the norm as much as it is now, so they probably had to endure a lot of folks trying to impart their own beliefs about their cultural differences just to be together and raise a family."

Those early experiences contributed to Nse's thinking about racial and environmental justice as she got older. In recent years, with Nse at its helm, CEHN has started taking a broader, more holistic approach to advocacy for children. Throughout the pandemic the organization highlighted that low-income children in communities of color, who faced the highest COVID-19 risk by virtue of having frontline workers as caregivers, faced additional threats from chemical exposures through poor housing conditions and old childcare and school buildings. The group has also begun pushing for action on climate change, noting that wildfires, extreme weather, and climate migration all threaten children's health, and that poor children and communities of color bear a disproportionate amount of the risk.

"Whenever we have the opportunity during a presentation, panel, or summit, we've called out that a Band-Aid approach to just one problem with children's health isn't going to work," Nse said. "If we don't address these problems at a systems-change level, they're just going to keep cropping up."

When she was in middle school, Nse decided she wanted to be a doctor. Specifically, a pediatrician. She'd been babysitting since she was thirteen—first her younger brother, then other neighborhood kids and relatives—and she admired her own pediatrician, Dr. Leslie Clapp. When Nse told Dr. Clapp she too wanted to help kids, the pediatrician offered to mentor her. Dr. Clapp encouraged Nse when, as a college student, she enrolled in a summer internship program with the NIH and won a leadership award while helping to research a rare respiratory illness at the Children's Hospital of Buffalo. "She was a busy woman, a pediatrician and a mother, but she always made time for me," Nse said. "I wanted to be just like her. I was like a mini Dr. Clapp."

Nse attended Siena College in Loudonville, New York, about four and a half hours from home, majoring in biology premed. But something shifted during her junior year when she was preparing to take the MCAT (Medical College Admission Test). Her test prep and premed courses felt highly technical and disconnected from the work she really wanted to do: helping children. If medical school and being a doctor were going to be like this, she thought, was it really what she wanted? "Everything I was doing was about this career, and I was afraid I was going to hate it," she said. "So I started asking myself if this was the only way I could go about achieving my goals."

She consulted with her college adviser, who "looked like a deer in the headlights" when Nse explained her dilemma. He didn't want her to waste the investment she'd already made or have to start over, so he suggested she think about becoming an osteopathic physician or a physician's assistant. But Nse knew those options wouldn't solve her problem. "He just wasn't hearing me," she said. "It was the first time I'd ever felt like I didn't know what to do with my life. It was a really high-anxiety time for me."

It didn't help that her parents were distraught over her sudden change of heart, and worried about what she would do for a living and all the time and money she'd already invested in becoming a doctor. Frustrated and unsure of what to do, Nse opted to finish out her premed degree. One of Nse's close friends, Jennifer Roberts, was having a similar experience of disenchantment with premed—only she was further along in the process when she started to panic. She had applied for a few med school programs and even gone to a few interviews when she realized it wasn't what she actually wanted. In 1997 the two of them sat in front of the computer in her friend's living room for hours, researching alternative career paths, and they discovered the field of public health. They both applied and were accepted to a few programs. Nse chose the public health program at George Washington University in Washington, DC, and Jennifer chose a different school and also went on to have a successful career in public health.

For eight months before starting grad school, Nse worked two jobs to save money for the move to DC. Her grandmother openly fretted about her granddaughter living alone for the first time in a big, crime-filled city, and Nse wondered whether she had made the right choice. She was terrified that she would get there and realize she'd messed everything up and ruined her chances of having a successful career that she cared about.

"But then, on what was literally my first day of class, I felt like I was being fully engaged in my education for the first time," she said. "I'd always felt impatient to get to the part where I could interact with actual people, and that was where that class started—with people, not with technical skills. Right away I knew I'd made the right choice. Biology and chemistry might teach you how to approach an individual problem for an individual patient, but that patient and doctor could do everything right, and if the broader community they live in isn't healthy, it might not really help. I became really excited about the notion of public health—everyone working toward a shared goal of creating healthy communities."

She went on to receive a master's degree in public health with a focus on maternal and child health, and right after finishing school, Nse took a job as a communications outreach specialist at the Children's Environmental Health Network. Within a year she became deputy director, then executive director soon after.

"I never would have guessed that I'd stay with one employer for two decades, but I have been so humbled and honored to do so," she said. "I learn and grow every day I'm there, and I'll stay as long as I continue to feel useful to the health and well-being of our most vulnerable."

————

From November to February 2020, while the pandemic raged on, Berry underwent chemotherapy every three weeks. The night before each treatment she had to take steroids, which often kept her wide awake. After each treatment, she would continue the steroid regimen for two more days, leaving her in what she referred to as a "faux energized state," but after the steroids had run their course, she would feel so nauseous and exhausted that she could barely move. She would spend the next five to seven days glued to her bed or the couch, then feel a little better for a week or two—keeping some food down, doing gentle yoga, and even venturing out of the house occasionally for a short walk or a massage or acupuncture appointment—just in time for the process to start all over again.

"It's hard to describe what chemo does," she said. "It literally poisons you. You feel like you've been poisoned. You're exhausted and sore and drained in a way you've never felt before. Even your senses get distorted. Smells become painfully acute. I'd walk outside and a car would drive by and it tasted and smelled like I was sucking on the exhaust pipe. Perfume was insufferable. They tell you to not even wear scented deodorant in the oncology department at the hospital because people undergoing chemo are so sensitive to smells."

Undergoing chemo during a global pandemic added an extra layer of misery. Berry kept some good friends in her "pod," but chemo suppressed her immune system, so she had to be particularly careful about exposure to the virus. That often meant isolation. "It was extremely lonely," she said. "I started to read these books about how to cope with having cancer and the first thing they all say is, 'Go hug your friends and family.' It was like, *fuck you, what am I supposed to do if I can't?*"

About two weeks after her first chemotherapy treatment, Berry's hair started falling out. A hairdresser friend cut her hair short, hoping she might not lose it all, but Berry still couldn't stand feeling clumps of it come out in her hands whenever she touched it, so she buzzed it off herself. "It was such a relief," she said. "I didn't care about how I looked."

Meanwhile, her relationship with her boyfriend started to fall apart. He had tried to be supportive. In the beginning he drove her to most appointments, and his mom even came to town for a while to help take care of Berry—but when she was there, he suddenly stopped being around as much, and it started to seem like he had actually called his mom in to do the hard parts so he could take a break. "I'd have to beg him to come spend time with me," Berry said. "It was really hurtful."

First he wasn't comfortable helping Berry with her fertility shots, then he wasn't comfortable helping her shave her head. He seemed increasingly stressed out by her illness. His dad had also recently been diagnosed with cancer, Berry said, so she tried to have compassion for the feelings her sickness was bringing up for him. Then one day when he was picking her up after a chemo session and she was feeling particularly awful, he shouted at her about something trivial. "He was so stressed and he refused to go to therapy, so he was not dealing with his emotions in a healthy way. He was taking it out on me—which is just not what you do," Berry said.

She realized she didn't have enough spare energy to fight with him or take care of him, so she ended the relationship in December 2020. They stayed good friends, and she knew their breakup was for the best, but that didn't make it any easier. Her anxiety and her loneliness were compounded by the loss.

———

Much of Nse's work involves building relationships, which means she knows a lot of people, including many who have had cancer. Three influential women in her life have had the disease. Two survived and one didn't.

Carol Stroebel worked at CEHN from 1996 to 2014. In 2015, during a prolonged battle with bile duct cancer, Carol received the organization's annual Child Health Advocate Award in Policy. She died the following year.

"I learned so much from Carol's tough love," Nse said. "She always said exactly what she thought, and she was a champion for advocacy through really understanding all the nitty gritty details of health policy. It was very difficult losing her."

One way that Nse coped was channeling her grief into work. She began representing the Children's Environmental Health Network in the Cancer Free Economy Network. When she was new in the role, the work changed the way she thought about cancer and cancer prevention. Before the pandemic, the group spent an entire year creating elaborate systems maps on wall-sized sheets of paper to identify where people are being exposed to cancer-causing chemicals, what economic and social systems drive those exposures, and where they could most effectively push for changes aimed at preventing as many cancer cases as possible.

"It's really rare to see such a comprehensive approach to this kind of work that includes everything from PFAS and pesticides to asbestos, formaldehyde, fossil fuels, and climate change," Nse said. "It made me realize that if advocates in other sectors took a similarly collaborative, systems-based approach, we could start making real progress."

"That's not to say it isn't complex," she added. "We know we won't be able to just flip a switch. It's going to be a long process, but I'm so excited to be part of it."

Through the Cancer Free Economy Network, Nse worked with nonprofit, business, and academic organizations in 2020 to publish a report that summarized all of the existing peer-reviewed science on environmental exposures and childhood cancer risk—marking the first time such an undertaking had been attempted, according to Nse.

For the report, *Childhood Cancer: Cross-Sector Strategies for Prevention*, Nse and her colleagues identified three categories of chemicals with robust scientific evidence of links between exposure and childhood cancers: pesticides, traffic-related air pollution, and paints/solvents. They also looked at chemicals that are carcinogenic to adults and prevalent in schools and childcare facilities, noting that while few studies have directly linked these chemicals to childhood cancers, "this absence of evidence reflects the difficulty of studying environmental contributors to rare diseases; it does not mean that carcinogenic exposures are safe for children." They addressed environmental justice, documenting the ways poverty and racism lead to additional risks and disproportionate exposures to hazardous chemicals for kids who live near industrial manufacturing, agricultural facilities, major transportation routes, and hazardous waste sites. The report also lays out the social and economic costs of rising childhood cancer, then reviews the incentives for reducing exposures.

For example, from 2000 to 2005, costs from hospitalizations related to childhood cancer in the United States doubled, and in 2009 alone, costs totaled $1.9 billion. A survey conducted by the National Children's Cancer Society found that one in five families with a child who receives a new cancer diagnosis are already living in poverty, and that families reported losing more than 40% of their annual household income because their work was disrupted during their children's cancer treatments, a figure that doesn't account for out-of-pocket expenses like traveling to the hospital or childcare for other kids at home. The report concludes that while "individuals can make choices that reduce children's exposures . . . a dramatic and equitable transition away from hazardous chemicals to safer alternatives, at the scale needed, requires action by businesses, community institutions, and government," and provides concrete suggestions on where to start.

The Network also provides a level of peer support that's rare in the world of NGOs and nonprofits. They launched a mentoring program to help newer advocates, in particular young people of color, develop into senior and leadership roles. And during the pandemic, they created a system for members to check in with each other, which resulted in the entire collaborative temporarily putting aside their usual work and using network funds to help with payroll and rent when member organizations were struggling to stay afloat.

"We're all part of this network because we believe the whole of our collective blood, sweat, and tears has the potential to fill in a gap that no one else is addressing," Nse said. "We all felt that caring for our members during that crisis had to be our first priority, because if we can't even take care of our own collaborative, how could we expect to take care of anyone else? It's not a collaborative anymore if you fall short in that regard."

In addition to being buoyed by her work with the Cancer Free Economy Network, Nse sees lots of other reasons to stay hopeful.

On March 11, 2021, Michael Regan was sworn in as the new EPA administrator, becoming the first Black man and only the second person of color ever to lead the organization. That day, Nse was among a small group of people representing NGOs and faith-based groups who were invited to meet with him. "I've been doing this for twenty years, and never before have I been invited to meet an EPA administrator on the day they were sworn in," she said.

The meeting was brief, but Regan listened carefully to Nse's input about the current state of children's health and her list of actions the federal government should be taking. Regan's first son died of neuroblastoma, a rare childhood cancer, when he was just fifteen months old, and his second son was still little, just seven years old, the day he was sworn in.

"He shared that keeping his seven-year-old son safe is a huge motivator for him in this work," Nse said. "He mentioned children and children's health at least as much as I did in the broader conversation, and I got the sense that he'll be a science-focused, forward-looking EPA administrator who is willing to work with organizations outside the government in order to get things done."

Five months later, that promise started coming to fruition: In August 2021, the EPA announced that it would ban the use of the pesticide chlorpyrifos on food crops—the same pesticide Nse had been trying to get banned in the state of Maryland before the pandemic. The change followed at least fourteen years of pressure from environmental, labor, and health advocacy groups like CEHN, and a long series of advances and setbacks. The EPA under the Obama administration had formally concluded that the pesticide couldn't be deemed safe and started working toward a ban, but those efforts were rolled back in the early days of the Trump administration.

"The average American has no clue how long it takes for these kinds of successes to happen," Nse said. "Even at the state level they involve so many legal experts, agricultural experts, and farmers, in addition to affected families, and so many weekly planning calls that lead up to those public testimonies, all of which are only a small part of the equation. These federal pushes involve even more large-scale coordination. Changes like this really are a huge win for everyone involved—and especially for children."

Another piece of federal legislation that Nse and other environmental health advocates have spent decades working on is the Toxic Substances Control Act (TSCA, often referred to as Tosca). The bill, which was originally passed in 1976, aimed to ensure the safety of new chemicals introduced to the market, but chemicals that were already on the market were exempted—many without having ever been tested or proven safe. The law also lacked teeth, ultimately resulting in court rulings declaring that the EPA didn't have enough authority to meaningfully regulate cancer-causing substances like asbestos under the law (as discussed in greater detail in chapter 2). In 2016, EPA administrator Gina McCarthy said TSCA was well intentioned, but "without major changes to the law, EPA couldn't take the actions necessary to protect people from toxic chemicals."[5]

TSCA got an update with the Lautenberg Act in 2016, which gave the EPA clear and enforceable deadlines for evaluating the safety of chemicals that are already on the market, strengthened the standards for assessing new chemicals, and expanded the agency's authority to request information about how chemicals are made. But some environmental and health advocacy groups were critical of the amendment. The Environmental Working Group pointed out that under the new law, states lost the authority to act more quickly than the EPA in stopping potentially dangerous products from going to market, and said that in order for EPA's new authority to be meaningful, Congress would need to provide a great deal of additional funding to the organization.

"The processes currently in place for evaluating the safety of products put on the market in the US are not adequate to protect children," Nse said. "New chemicals still aren't being adequately tested for their impact on human health and development. Companies are still churning out thousands of new chemicals each year, and only a handful of them have been thoroughly tested for human toxicity."

Most of CEHN's work related to TSCA has happened in collaboration with other public health and environmental advocacy groups. "We've learned that we have a much stronger impact when lots of diverse groups are all singing the same tune," Nse said. "Though the pace of progress continues to be frustratingly slow, it gives me hope seeing these groups come together and realize we're so much more effective when we leverage our collective power."

In May 2022, Nse discussed environmental chemicals on a panel convened by the Cancer Moonshot, a federal program aimed at accelerating progress on the disease. That event and the federal chlorpyrifos ban were both major milestones, but Nse knows better than to pin all of her hopes on one presidential administration. "We'll see what happens next," she said. "We plan to be very direct about what needs to happen to create meaningful, lasting change for children's health in this country, and hold these new leaders accountable. It's going to take some time, but I think we're in a very unique situation that makes change possible right now."

CHAPTER 4

Bill: Safer Homes and Offices through Market Pressure

Bill Walsh was born in Salem, Massachusetts, in 1959, on the eve of a decade rife with political and social change.

His earliest memory is waking up and hearing the news of John F. Kennedy's funeral on television. Salem is a college town, and he remembers the protests of the civil rights, labor, and antiwar movements as the formative milieu of his childhood. That probably has something to do with his unusual career path—from lawyer, to activist, to founder of a highly effective nonprofit, to head of a grantmaking foundation.

More than twenty years ago, Bill founded the Healthy Building Network (HBN), an organization dedicated to reducing toxic chemicals in building materials.

We tend to think of air pollution as something that affects us when we're outdoors, but concentrations of air pollutants are often two to five times higher indoors, where the average American spends 90% of their time. Building materials like floor tiles, carpeting, upholstery, insulation, paint, and treated wood often contain potentially cancer-causing chemicals like asbestos fibers, phthalates, polybrominated diphenyl ethers (PBDEs), formaldehyde, per- and polyfluoroalkyl substances (PFAS), short-chain chlorinated paraffins (SCCPs), and chromated copper arsenate (CCA). Chemicals in building materials can off-gas over time and be inhaled by occupants. Exposure also happens through physical contact with building surfaces and dust from building materials, especially for babies and small children, who explore with their hands and mouths.

These chemicals are especially harmful for the people in manufacturing, construction, and facility management whose jobs require constant exposures to these materials. And they continue increasing people's cancer risk after they are disposed of—either by leaching chemicals into groundwater, creating hazardous emissions if they're burned, or continuing to off-gas as they break down, which can take hundreds or even thousands of years.

Bill, who has curly salt-and-pepper hair, a gray beard, and kind eyes, grew up in a working-class household. Neither of his parents had college degrees, and his dad worked as a milkman, for a railroad, and as a salesman at various points, often working second and third jobs on evenings and weekends to make ends meet while his mom stayed home to care for Bill and his two younger siblings.

Bill's dad was also active in the local Democratic Party, helping run local campaigns. When Bill was twelve, his dad ran for a position as ward councilor. Bill remembers helping him knock on nearly every door in the ward over the course of a summer. There was a precinct filled with Spanish speakers that white locals referred to as the Puerto Rican section of town, but Bill learned that actually, most of its residents hailed from the Dominican Republic. His dad's campaign managers told him not to bother going there—that it wasn't safe, that there would be a language barrier, and that there weren't many votes to be gained from it anyway.

"He went anyway, despite not speaking a word of Spanish," Bill said. "He had his written materials translated into Spanish, and I remember him explaining to me that if you want to get elected you always have to respectfully ask for people's votes, and that if he won he would represent this neighborhood, so it wouldn't be right to ignore them." This was just a small lesson in basic civility, Bill said, but it stuck with him.

His mother, meanwhile, despised politics. "She hated the whole thing," Bill said. "To her credit she smiled in all the family photos in the brochure and she showed up to all the events, but at home she was very clear that she was not looking forward to being a political spouse. While we were out knocking on doors she used to cut out these little comics from the local paper and tape them on the fridge—I remember one that said, 'To err is human, to blame it on someone else is politics.'"

Helping with his dad's campaign was exciting and sparked his interest in politics, but his mom's attitude kept his perspective on the whole endeavor grounded.

"In the end he lost," Bill said, "which I think my mom was relieved about, though she graciously never said that out loud."

Bill's dad also taught him the importance of caring for the environment. Salem had once been full of leather factories that dumped animal and chemical waste into the North River, which runs through the town. When the tide receded, the stench that came off the river was nearly unbearable, and Bill's dad joined a local group working to get it cleaned up. Progress was slow, so they turned to a creative form of protest. Every summer the town held a big parade to celebrate its heritage. Bill's dad helped build a huge, hideous, papier-mâché fly with a bobbing head. They put it on a flatbed truck and crashed the parade route, donning fly costumes and "swarming" around it while handing out flyers requesting support for their cleanup efforts.

Throughout his thirties, Bill worked at Greenpeace. When he told this story to his colleagues there, they joked that it had been his first "direct action," a precursor to a career with an organization known for strikes, tree sitting, lock-ins, and banner dropping.

"I hadn't thought of it that way," he said. "My parents had pretty mainstream democratic views. They never took me to an antiwar or civil rights protest. But I guess I was exposed to that kind of political action pretty early in life."

————

Berry's grandmother on her dad's side—the one she's named after—came from oil money. Her father (Berry's great-grandfather) was one of the cofounders of what eventually became Quaker State oil company.

Berry's grandparents' old house, which has now become the family hub, is a 10,000-square-foot, three-story Tudor on 9 acres of land in Oil City. Berry showed me the house during our tour of Oil City in December 2021. It was built by the founders of Pennzoil at the height of the oil boom in the 1920s, and the Breenes bought it from their descendants in the 1960s. A long horseshoe driveway curves through old-growth trees to make a loop around the front of the house, which can only be described as castle-like. A mix of brick and intricate stone-work frames the arched doorways, tall windows, and a large turret. The driveway branches off through a tall stone archway, leading to a four-car garage that's attached to the back of the house beside the kitchen door, which the family uses as their main entrance. The yard is flanked with rhododendron bushes, once carefully cultivated and tended by Berry's grandmother but now so overgrown and spread out that each resembles a small forest. There are various side patios with little awnings and collections of tables and chairs, and aging pillars that once flanked a fountain overlooking acres and acres of hilly woods.

A few of the rooms have been redecorated over the years—including the kitchen, which looks like it could have been lifted from the set of *The Brady Bunch*, with its plaid blue and green wallpaper, and a few upstairs bedrooms and bathrooms decked out with the fuchsia-dominant floral patterns of the 1980s—but for the most part, the house pays homage to the opulence of early 1900s Pennsylvania oil money. Palatial sitting rooms are outfitted with plush rugs, antique velvet love seats and wingback chairs, classical paintings with brass lights mounted above their top frames, and sprays of decadent silk flowers rising from ornamental porcelain vases. There are colorful murals painted by a local artist on the walls in the main dining room, and bone china sets fill a dozen cabinets and sideboards. Framed family photos, candelabras, and antique curios crowd the tops of various bonheurs du jour and chiffoniers—silver lighters, baroque lamps, brass hairbrushes and clocks, porcelain and jade figurines acquired on long-ago voyages. There's an original working elevator, a baby grand piano, a billiards room, a servant's quarters, and a spiral staircase up the turret.

In the dining room, Berry showed me what had been her favorite feature of the home when she was a kid: two curved, inset curio shelves flanking the mantelpiece in the dining room that double as secret doors. Flip a switch beneath the mantle, and after some mechanical clicking the shelves unlock and can be swung outward to reveal two small cream-colored cabinets filled with built-in shelves and drawers that were likely used to hide alcohol during prohibition. Now they're stocked with Berry's grandparents' spare linens, napkin rings, and opulent serving ware.

Berry's family has been involved in local politics for as long as she can remember. Her grandparents and parents were all prominent figures in the local Republican Party, serving various roles through the years, and her brother is currently a county commissioner. He lives in the house now, as do Berry's half-sister and her half-sister's husband and daughter. There's so much space that Berry could move in too, if she wanted, and the family would rarely even have to encounter each other in the house—but despite the difficulty of the last few years, Berry prefers her independence and her little house at the top of Troy Hill in Pittsburgh.

At the end of the street that leads to the house, there's a little turnaround with an overlook onto a picturesque view of rolling hills flanking the valley cut through by Oil Creek. From here you can see Oil City below and, to the right, the adjacent borough of Rouseville.

Rouseville was named for Henry Rouse, a former teacher and legislator who struck it rich after discovering a "gusher" in the earliest days of the American oil industry. Rouse's oil well was so productive it couldn't be contained—3,000 barrels of oil a day shot up toward the sky, coating everything nearby in oil that eventually caught fire and exploded, killing Rouse and nineteen others on April 17, 1861. Rouseville rose from the ashes to become a hub of oil refineries. They're all gone today, replaced by other industrial facilities, but several refineries were still in operation while Berry was growing up, and she can viscerally recall the awful smell that rose from their smokestacks and singed the air in the valley.

The extraction, refining, and burning of fossil fuels are major sources of exposure to cancer-causing chemicals, such as benzene, dioxins, and polyaromatic hydrocarbons. Berry doesn't remember it, but her brother told her locals used to refer to Rouseville as "cancer valley" because of the high number of cancer cases among residents.

———

When the Healthy Building Network was founded in 2000, one of its first initiatives was tackling the use of polyvinyl chloride (PVC) plastic, also known as vinyl. PVC is cheap and durable, so it's often used in flooring, roofing, siding, windows, and doors. But PVC is also considered one of the most toxic plastics on the market. Vinyl chloride, the chemical used to make PVC, is a carcinogen associated with increased risk of a rare form of liver cancer (hepatic angiosarcoma), as well as brain and lung cancers, lymphoma, and leukemia.[1]

At the time, the US Green Building Council (another nonprofit) was developing the Leadership in Energy and Environmental Design (LEED) certification system, which would eventually become the most widely used green building certification in the world. The lead author had proposed issuing LEED credits to buildings that didn't use PVC, but the plastics industry was pushing back against that idea. HBN filed a brief and defended the provision in front of the board of the US Green Building Council, but the proposed credits were not adopted.

"That effort was halfway successful," Bill said, "because the decision makers did publish a report that concluded PVC was 'consistently among the worst materials for human health impacts,' and as a result many industry leaders did shun PVC." But the group wouldn't completely rule out the use of PVC because alternative materials weren't always environmentally superior, and some products, such as vinyl windows, had energy efficiency benefits.

"To this day most green building certifications remain heavily skewed toward energy efficiency," Bill said. "To the extent that they do include credits for using nontoxic building materials, they're often not worth enough points to really incentivize the use of healthy materials."

Around the same time, Bill learned that due to a loophole in EPA regulations, most pressure-treated wood used for residential decks, picnic tables, gazebos, playgrounds, and fencing was still treated with the arsenic-based pesticide CCA. A very high dose of arsenic can kill you, but prolonged exposure to lower levels of the poison can cause a variety of illnesses, including skin cancer, bladder cancer, and lung cancer.

"If the amount of arsenic in a pressure-treated two-by-four was collected in a bag, it would be illegal to dispose of it in a landfill," he said, "but you could legally throw out that two-by-four anywhere and it would still leach the same amount of arsenic into the environment."

The manufacturing process for CCA-treated lumber endangered factory and construction workers who handled it. Numerous carpenters got severe, acute arsenic poisoning from working with pressure-treated lumber or burning wood treated with CCA. There were reports of horses and dogs dying after chewing CCA-treated wood, of people losing fingers after getting splinters from CCA-treated wood, and of workers who'd spent decades handling CCA-treated wood being diagnosed with cancers associated with arsenic exposure.[2] At the time, stores like Home

Depot and Lowes sold picnic tables made from wood treated with CCA, typically with a label warning that the wood should never come into contact with food—something most consumers were unlikely to remember during a picnic.

This seemed like an obvious target, but after the lessons they'd learned from their PVC LEED campaign, HBN launched a comprehensive investigation to determine whether there were equally effective alternatives to arsenic. They learned that not only did safe alternatives exist, but they had also already been thoroughly tested and proven to perform equally well. "We felt that we had a real shot at getting arsenic out of pressure-treated wood since there was this ready-made alternative that was effective and a comparable price, CCA was well known to be harmful, and so many uses of arsenic had already effectively been banned otherwise," Bill said.

HBN partnered with other organizations working toward the same goal, including Clean Water Action and the Environmental Working Group. After about a year of campaigning and giving scientific testimony, they were successful: the regulatory loophole was effectively closed in 2003 through a voluntary phase-out agreement between pressure-treated wood manufacturers and the EPA. Those efforts ultimately reduced the use of industrial arsenic in the US from more than 20 metric tons a year to about 6 metric tons a year, according to an analysis by HBN.[3]

Other harmful building materials have proved more difficult to tackle. In the early 2000s, many companies realized there were economic benefits to marketing their products as "green," so they began offering more energy-efficient alternatives where it was easy to do so. "Whether they sell a white roof or a black roof, they're still going to sell a roof," Bill said. But when it came to replacing toxic chemicals in their products, the companies that manufactured building materials were less cooperative.

"Manufacturers might fight over the timeline for implementing new energy efficiencies, but they basically agree with the premise that more energy efficiency is good," Bill said. "But nobody wants to see that their products are not on the list of healthy building materials, which becomes very contentious. The industry fought very hard against any intrusion into business as usual, and would often adamantly insist that whatever they were doing was 'green' despite clear evidence to the contrary when it came to protecting people's health."

There was a learning curve to figuring out how to tackle these more complex problems. When Bill first started the Healthy Building Network, he said, he had lots of ideas about how to make building materials less toxic, but running a nonprofit was brand new to him since he's a lawyer by training.

In high school, Bill wrote a term paper on Dee Brown's book *Bury My Heart at Wounded Knee: An Indian History of the American West*, which planted a seed in his mind. "Here I was, fifteen years old, realizing I'd never heard this history despite taking lots of history classes," he said.

He went to the library and was shocked to find that it wasn't because that history was hard to come by. Using a microfiche machine, he found old feature stories in the *New York Times* about groups of native tribes and white allies fighting for fair treaties. "That opened my eyes to the fact that there are always multiple sides to the story, and you often have to dig deeper to find them," he said.

Bill lived at home and worked various food service jobs while attending Salem State University—McDonald's cashier, bartender, waiter, and cook. He was an English major with no idea what he wanted to do for a career until one of his professors suggested law school. He passed the LSAT, got into a few schools, and chose Northeastern University because of their co-ops—a program where law students alternate semesters in the classroom with semesters spent working full-time legal jobs. "At the time I was primarily focused on getting a nice car," he said. "I thought that program would let me work my way through law school, come out a lawyer and get a Porsche."

Things changed once he started his studies. Many of his classmates were older people pursuing law degrees as a second career. He fell in with that older crowd, some of whom had worked with labor unions and were becoming lawyers to be more effective in that work. "Spending time with them calmed me down and convinced me to make the most of my time in law school," Bill said.

Bill's older classmates told him he would pass the bar exam as long as he got enough real-world experience, so instead of taking basic bar courses he opted for a course load heavy on clinical studies. As a student he represented prisoners in maximum security units at internal disciplinary hearings, worked at a housing rights legal clinic, worked for the American Civil Liberties Union, and worked at a labor law firm—none of which paid more than a stipend, but all of which he found incredibly interesting and fulfilling. "I abandoned all hopes for a Porsche," Bill said, laughing, "but making these choices to follow my heart I met all of these fascinating, educated, brilliant people who showed me that I didn't have to have a boring legal career."

During the semester that he worked at a labor law firm in Arkansas, Bill was mentored by an attorney who would attend depositions and negotiations in a short-sleeved shirt with no tie. When Bill asked why, his mentor said he cared more about the union members he represented feeling comfortable with him and trusting him than he did about looking like a lawyer on a superficial level, which resonated deeply with Bill. Another semester, he got to work with the late A. Leon Higginbotham Jr., the first Black judge ever appointed in the District Court for Eastern Pennsylvania, who later served as chief judge of the US Third Circuit Court of Appeals and was awarded the Presidential Medal of Freedom by Bill Clinton.

After finishing his law degree, Bill pursued a master of laws in public interest representation at Georgetown, which is where he started thinking about environmental issues. One of his cases involved a woman who got sick every time a lawn service company sprayed herbicides in her neighborhood. She had asked the company to let her know before they came, but the company had refused. So the woman convinced her local county counsel to pass a regulation requiring that residents be notified before the herbicide treatments. The lawn care company fought back, claiming that any local attempts to regulate this type of chemical were preempted by federal law under the Federal Insecticide, Fungicide, and Rodenticide Act.

"That's where I was assigned to the case," Bill said. "It didn't seem like an environmental issue to me as much as a justice issue at the time, even though our client was an upper-middle-class white woman so it wasn't an issue of race or class. I was just struck by the arrogance of the company, this refusal to accommodate a simple request. It was a revelation learning about preemption and how this federal law made it really hard for any locality to protect its residents from toxic chemicals."

While working on her case, Bill met people at the Rachel Carson Council, a Maryland-based organization doing advocacy work related to pesticide use, and at the National Coalition Against the Misuse of Pesticides (now called Beyond Pesticides). He also connected with people at the Center for Health, Environment and Justice, a national organization founded by Lois Gibbs.

After noticing that her community experienced high rates of birth defects, miscarriages, cancer, epilepsy, asthma, and kidney disease in the 1970s, Gibbs discovered that her children's elementary school in Niagara Falls, New York, was built on top of a 21,000-ton toxic waste dump formerly owned by the Hooker Chemical Company, which had sold it to the school district without doing any cleanup. Gibbs was a housewife with two young children and no prior experience in activism, but she organized her neighbors and led a years-long battle that eventually led to the evacuation of more than 800 families and an extensive cleanup of the Love Canal in 1978. Ultimately, Gibbs's work led to the creation of the US Environmental Protection Agency's Superfund program, which is still used to locate and clean up toxic sites today.

"I found Lois Gibbs's story to be just shattering," Bill said. "It made me realize these people are fighting not just for a cleaner environment, but for their lives and their kids' lives."

Bill was interested in representing a union, but when he was finishing his fellowship in the mid-1980s the Reagan administration undermined that plan. Unions across the country were being busted at the time, and there were very few legal career opportunities.

"I saw this environmental thing as another fight for justice," he said, "though at the time I didn't anticipate how big it would get, or what an all-consuming part of my life it would become."

When he finished graduate school, Bill accepted a position with the US Public Interest Research Group (PIRG), a federation of nonprofit organizations dedicated to consumer protection and public health. He became a Superfund attorney, tasked with testifying before Congress and overseeing implementation of the law. But the work wasn't quite as fulfilling as he'd hoped it would be.

"I found that work really frustrating," he said. "It was very slow-going and hard to get a truly satisfactory outcome. There were endless debates about what to do with these vast amounts of terribly contaminated soil and about how clean is clean enough, and I felt like citizens were always losing. The prospect of getting ahead of all that and reducing the flow of toxic chemicals to begin with was really appealing to me."

Through his work with US PIRG, Bill connected with Massachusetts PIRG, which was working with other academic and environmental advocacy organizations toward what they had termed toxics use reduction: eliminating pollution at the source instead of generating toxic waste dumps that would later have to be cleaned up. Their work would eventually culminate in Massachusetts enacting the Toxics Use Reduction Act (TURA) in 1989. The law requires companies to track, document, and report their use and disposal of certain toxic chemicals, and to make a plan for reducing their use over time. It's a unique approach because the planning requirement doesn't actually mandate a measurable, specific reduction of toxic chemicals, but the act of creating a plan to reduce their use often reveals inefficiencies and presents opportunities to create safer processes.

After TURA was passed, the use of cancer-causing chemicals declined by 32% in Massachusetts over the next two decades, and releases of known or suspected carcinogens into the environment declined by a whopping 93%.[4] TURA remains one of the most effective state laws regulating toxic and cancer-causing chemicals in the US, and it has also been recognized internationally and used as a model for other countries.

Greenpeace was part of the coalition working on toxics use reduction strategies in Massachusetts, so Bill got to know them too, and when a legislative director position at Greenpeace opened up, he decided to apply. "In many regards it should have been an unappealing position for a Washington lawyer," he said. "Greenpeace was considered too radical by many other environmental groups, especially those working on legislation changes on Capitol Hill. But I liked that they had a broader range of tools than PIRG or other lobbying groups through their use of direct action and grassroots organizing tactics."

After about a year working as Greenpeace's legislative director, he shifted into a position as the organization's campaign director—a job that rendered his law degree less relevant, but that he found very fulfilling. During his law school co-ops, Bill had learned that there are two kinds of lawyers who dedicate their careers to helping people.

"Some people couldn't live without the satisfaction of representing a specific client, and they felt so good about what they could do for that one person," he said. "I was the other kind. I felt terrible knowing we represented so few clients while there were so many more outside the door who were never able to get the legal services they needed. People getting evicted, prisoners losing months of freedom. And half the time even when someone was getting free legal help, their case would be completely hopeless. I got depressed by that. I felt like I couldn't keep looking these people in the eye knowing I couldn't actually help them."

"When I started to shift into policy work I had friends who were in that first camp who said, 'Why would you want to be a lawyer if you never even get to see the people you help?' But I learned that for my own well-being, I felt better working toward policy changes that could help thousands of people at once, even if I'd never get to meet them, than I did trying to help one person at a time under what were often impossible conditions."

At Greenpeace, Bill was able to work both on policy and directly with communities, which felt like the best of both worlds. He got to engage with grassroots community struggles and push beyond dealing with toxic waste to find ways of reducing it. It was satisfying work. He kept the job for a decade, and it was how he met Jennifer Carr, his partner of more than 25 years, who worked in the fundraising department at Greenpeace at the time and also still does work related to the environment. They have two children and a granddaughter together.

Bill's predecessor at Greenpeace had started working with grassroots groups across the US at what Bill called the "frontlines of the struggle against chemical pollution." Bill inherited that legacy and ran with it, but this was very new terrain for the organization and required different tactics than some Greenpeace organizers were used to.

"It's very different just swooping in and hanging a banner to bring awareness to an issue—or even being out on the open seas blocking a Japanese whaler ship—than it is being in a small town in Ohio trying to stop a toxic waste incinerator from being built," Bill said. "We realized that without building relationships that enabled trust and coordination, we could easily undermine the efforts of local people who were already working on things and would have to live with the outcomes. Historically, Greenpeace was an independent agent that operated on conscience, so it was a big shift when we started working collaboratively with people in these communities."

Bill took over as the Greenpeace campaign director in the early 1990s, during what he refers to as "the first wave of the environmental justice movement," when activists of color started challenging mostly white, Washington, DC–based environmental advocacy groups to diversify their staff and defer to people fighting for change in their own communities.

"I thought that was really important to do, especially in toxics work," Bill said. During his tenure, the Greenpeace Toxic Campaign diversified its staff so that about 40% of program positions were held by people of color—most of them accomplished, heavy-hitting activists who had already been advocating for their own communities for years.

"Now we call it intersectionality and diversity, equity and inclusion," Bill said. "At the time we didn't have that language. We just knew we needed to build teams that were rooted in the communities being harmed in order to be effective."

And they were effective. They helped communities stop several toxic chemical manufacturing and disposal facilities from being built or expanded around the country, including petrochemical plants, which convert fossil fuels into plastics and emit a slew of cancer-causing chemicals. Many of the community activists Bill helped recruit to Greenpeace went on to make indelible marks on the American environmental movement. Juan Parras, a Texas-born labor and environmental justice organizer, helped lead Greenpeace's successful fight against the Japanese petrochemical company Shintech in Louisiana, and later founded Texas Environmental Justice Advocacy Services. Parras was appointed to President Biden's Environmental Justice Advisory Council in 2021. The late Damu Smith, a racial and environmental justice organizer from

St. Louis who organized tours of toxic facilities in the South for Greenpeace, later coordinated one of the largest environmental justice conferences ever held, then cofounded and served as executive director of the National Black Environmental Justice Network. Jacquelynn "Jackie" Warledo, a Seminole Nation Tribal Citizen who served on Greenpeace's Indian Lands Toxics Program, cofounded and served as a National Council member of the Indigenous Environmental Network, and as a lead Indigenous negotiator for the development of the UN Stockholm Convention on Persistent Organic Pollutants before her death in 2019.

"Many of my colleagues were more experienced activists than me who had spent their whole lives engaged in serious civil rights struggles," Bill said. "Witnessing that kind of community organizing was a really profound experience that completely changed the way I thought about the environmental movement. To see how interlinked the issues of racism, poverty, and pollution were, and to discover the power of the activists and leaders in those communities made me realize that we weren't going to be able to 'save the environment' without addressing systemic racism—and that if you could address systemic racism, you'd also have a good shot at saving the Earth."

Unfortunately, not everyone in the organization was having the same revelations. At the time, Bill said, some white leaders at the international level thought focusing on American social issues was a distraction from their work on the environment, rather than recognizing the ways in which the movements were inherently intertwined, and Greenpeace began backing away from those communities and commitments.

"Today, environmental justice is central to a lot of big environmental organizations' work," Bill said. "But I think we were a bit ahead of our time in the way we thought about institutional, societal, and racial issues." He added that he still believes those leaders should have known better at the time, and that it was "terrible judgment to stop those programs."

In 1991, Bill was invited to attend the First National People of Color Environmental Leadership Summit as an observer. Participants from all fifty states, Puerto Rico, Chile, Mexico, and as far away as Nigeria and the Marshall Islands drafted the "principles of environmental justice." That meeting and those principles laid the groundwork for the first presidential executive order in the US aimed at protecting low-income communities and communities of color from the disproportionate effects of toxic pollution, which was signed by President Bill Clinton on February 11, 1994.

"I felt like I was bearing witness to history," Bill said. "Watching this group of multiracial, multicultural organizations navigate real organizing challenges and bring each group's voice to the dialogue was so beautiful and inspiring. Many of the people leading that conference are now sitting on the president's environmental justice council, and to see that those principles are back on a presidential agenda today is very satisfying."

As Bill's work on toxic exposures at Greenpeace evolved, so did his thinking about tackling problems one at a time versus systemically. After concluding a lengthy campaign related to lead exposure, Bill and his colleagues felt overwhelmed at the prospect of continuing to tackle harmful exposures one chemical at a time. They looked to their Dutch, German, and Swedish counterparts, who were pioneering the notion of regulating entire classes of chemicals, and decided to focus on the whole suite of halogenated compounds—a category that includes chlorofluorocarbons (CFCs), polychlorinated biphenyls (PCBs), and PVC. CFCs were burning a hole in the ozone layer, while PCBs and PVC were found to increase cancer risk.

The thinking was that if they could get the whole class of halogenated chemicals more strictly regulated or banned, they could simultaneously stop harmful exposures that occurred at every phase of their existence— from harmful emissions created during the manufacturing process, to causing acid rain and burning holes in the ozone during their use, to the near-impossibility of safely disposing of them at the end of their life cycle. Greenpeace campaigned heavily against CFCs from 1986 to 1995, helping drive an international treaty in 1987 that ultimately cut their use in half. Their efforts (along with those of many other advocacy groups) eventually resulted in 197 countries banning the use of CFCs, which has helped the ozone layer almost completely recover.[5]

PCBs were banned from being manufactured in the US in 1979 under the Toxic Substances Control Act, so Greenpeace's work during the 1980s and 1990s focused on stopping manufacturing overseas and cleaning up sites across the US that remained contaminated with the compounds. By working with chemists and toxicologists, Bill and his colleagues discovered that PVC plastics could be easily replaced by less toxic alternatives that were just as effective. They started pressuring companies to switch to those alternatives. Advocacy work aimed at reducing PVCs is ongoing today, but thanks to the work of Greenpeace and other activists, a number of large plastics manufacturers have switched to safer options, and consumer awareness about the dangers of PVC has grown tremendously.

"Sometimes environmentalists are accused of throwing darts and just not liking any chemicals, but all of our work was really strategic and aimed at reducing exposure to the most harmful, widespread chemicals for as many people as possible," Bill said. "When I saw how effective it could be, this kind of systems-based approach became foundational to my thinking."

———

After her cancer diagnosis, Berry found herself thinking more about her childhood and her family's history—both the sweet parts and the difficult parts.

Her parents, Charles and Martha (aka Marty), met on a blind date arranged by a friend. Charles was two full hours late—typical for her dad, Berry said—but Marty stayed because the bartender kept serving her free glasses of scotch out of pity that she was obviously being stood up. By the time Charles finally showed up, Marty was in too good a mood to stay angry. They hit it off and got married a few years later. It was the second marriage for both, and they were in their midthirties. Marty was pregnant with Berry's older brother Sam at the wedding, and Berry was born 15 months later, when Marty was thirty-nine years old.

Berry and Sam went to the Catholic elementary school a few blocks from their house. They were Episcopalian, but their parents thought the Catholic school could provide a better education than public school. It made the kids feel like odd ones out, which Berry handled by being quiet and shy at school—though she was outspoken and rambunctious at home—and which Sam handled by being a class clown and smart-mouth. They both switched to public school midyear when Berry was in seventh grade because her brother got suspended for reasons their parents disagreed with.

Berry's school years were packed with activities: the diving team, track team, cheer team, skiing and snowboarding, student council, theater, art club. She was a good student and had jobs babysitting and working at a local diner. At the end of her senior year of high school, Berry was voted best dressed, most artistic, and "most interesting individual" in her class's yearbook superlatives, but was told she could only officially win in one category. She chose "most interesting individual."

She remembers her childhood as mostly easygoing and happy, but the ordeal of having cancer also called to mind a series of traumas that swirled around her in Oil City.

In 1992, when Berry was eight years old and in second grade, an eleven-year-old girl named Shauna Howe was kidnapped and sexually assaulted while walking home from a Girl Scouts Halloween party, then thrown off a bridge to her death. The incident received national media attention, and Berry said it changed the feeling of her small hometown.

The Oil City Council voted to ban nighttime trick-or-treating, a policy that remained in effect until 2008. The perpetrator of the crime remained unknown, and most parents grew paranoid and protective, forbidding their kids from walking through town alone.

"It kind of set the tone for the town all during my childhood," Berry said.

The case remained unsolved for nearly a decade before advances in DNA technology identified a pair of local brothers who were convicted of the crime, which unsettled everyone even more—they had always assumed it was outsiders, strangers who'd come through town and committed the horrendous crime, but it turned out to be two of their own.

Around the same time as the murder Berry's brother Sam got hit by a truck while running across the street after a neighbor. The truck was moving slowly so he escaped major injuries—he just needed stitches in his forehead—but Berry remembers her mother answering the phone and shrieking, and that when they went to see him in the hospital his yellow T-shirt was covered in blood.

When she was in middle school, two girls in the grade ahead of Berry were killed by a train while putting pennies on the track during a family reunion. Another train was passing on tracks nearby, so they didn't hear the second train approaching. The whole town was traumatized again, the death of the young girls recalling Shauna Howe's violent murder, and Berry remembers their funeral being crowded with heartbroken mourners.

When Berry was sixteen, her brother's best friend died in a car accident, and she was the one to tell him. Throughout her high school years, as the opioid crisis surged in towns like Oil City across the country, Berry lost numerous friends to addiction, overdose, and imprisonment. "I feel like almost every year there was some tragic death," she said. "Car accidents. Overdose deaths. Suicides. I went to a lot of funerals."

When she was thirty years old, Berry also lost a friend to stage 4 breast cancer. "I thought I was pretty desensitized to death by then, but when Lindsay died it hit me pretty hard," she said. "I think because I realized that she hadn't made any choices or mistakes that led to her death. It just felt really, really sad."

––––––

In 1999, Bill witnessed two things that made him think about building materials in an entirely new way.

First, after years of campaigning by Greenpeace and other environmental advocacy groups like Rainforest Action Network and ForestEthics, Home Depot committed to stop buying wood from endangered forests and only purchase wood that had been third-party certified as sustainable by the Forest Stewardship Council.

"I knew Home Depot was a big company, but I hadn't realized what a global market driver they were," he said, explaining that Home Depot is one of the largest buyers of wood on the planet, along with Lowes and Ikea. "The quantities of forest acreage this change affected was just vast."

Once Home Depot had signed on, other large companies followed suit to stay competitive, amplifying the reach of those campaigns even further.

Around the same time, Greenpeace USA was beginning construction on its new office headquarters in Washington, DC, and Bill joined the committee of staffers who volunteered to meet with prospective architects to discuss green building materials.

"I honestly felt bad for the guy," he said. "I thought I'd be objecting to every piece of plastic and worried about toxics, and the forest person would be worried about every piece of wood, and the climate person would be worried about energy efficiency, and this poor architect would just end up feeling defeated."

But the architect had done his research. He hoped to build a practice around green buildings and thought that if he could satisfy Greenpeace it would firmly establish his credentials. He had already worked with a number of other environmental organizations, and, Bill said, "he had really learned his stuff."

"He was incredibly well versed on all of the issues we raised, and he was actually less siloed than we were. So between me and my colleagues focused on different issues, none of us knew that much about the others' domains, but here was this architect who could offer solutions and clearly explain the tradeoffs different choices would make across those different spheres."

The knowledgeability of the architect and the massive industry shift driven by Home Depot led to an epiphany for Bill. "I realized there was a huge change under way," he said. "I was already thinking about how to affect decisions about what materials get used at the beginning of a manufacturing process, rather than regulating bad behaviors and cleaning up negative outputs—and suddenly building materials seemed like a perfect place to start that work."

Bill's friend Gary Cohen had started a group called Health Care Without Harm, which had begun reaching out to hospital systems regarding the incineration of medical waste, which often happened in densely populated urban neighborhoods without much effort to control harmful emissions. Gary and his team found that hospital administrators were concerned about being good neighbors and not causing more of the health problems they treated, including cancer, and were generally receptive to the group's proposals. They formed the first environmental group for hospital systems, and later the EPA followed their lead. Today, the group still has a substantial impact on the materials used in healthcare and how they're disposed of. Bill proposed using a similar model for the building materials industry, and Gary thought it was a great idea. He helped Bill raise initial startup funding for the Healthy Building

Network. They set out to find someone to run the organization but struggled to find the right person. "Eventually he said, 'If you believe in this, I think you have to do it.'" Bill said. "So I decided I'd give it a try."

For the first few years, Bill ran HBN like a small family business. He recruited people he knew and trusted, mostly activists he'd met during his time at Greenpeace. Most of them didn't have any experience with the building industry, but they all had experience in effective environmental campaigns and thinking on their feet. "We started out with absolutely no idea what we were doing in the building industry," Bill said. "But we were able to quickly gain support and start figuring it out."

More than twenty years later, the organization is a powerhouse. Initially they focused on tackling worst-in-class chemicals one at a time. They had some early success with things like getting arsenic out of pressure-treated wood, but they were surprised by how aggressively the industry clung to other toxic substances. Bill pointed to the example of the conservative billionaire Koch brothers, who led an aggressive lobbying effort aimed at keeping the US Department of Health and Human Services from labeling formaldehyde, which is widely recognized to cause leukemia, as a carcinogen. Georgia-Pacific, a subsidiary of Koch Industries, is one of the world's largest formaldehyde producers. The agency officially classified formaldehyde as a "known carcinogen" in 2011.

"It's astounding how many people in this industry defend the use of formaldehyde, a known carcinogen, especially given the level of public support for anything that helps fight cancer," Bill said. "It can be harder to explain the dangers of something like endocrine-disrupting chemicals to people, but replacing cancer-causing chemicals should be a no-brainer. Yet instead of looking for alternatives to formaldehyde, they're still pushing regulators to reclassify formaldehyde as noncarcinogenic."

Formaldehyde is used in several ways in the buildings we inhabit. It's used to keep draperies and upholsteries wrinkle-free and stain-resistant, and to bond particleboard and plywood together. Formaldehyde can also be formed as a by-product when chemicals interact, so it can wind up in products that don't have formaldehyde added directly. Until 2015 formaldehyde was also used to hold together the fiberglass in batt insulation (material used between walls and the frame of a building). But thanks to efforts by HBN in collaboration with other advocacy groups, every manufacturer of residential fiberglass insulation in the United States and Canada voluntarily and completely phased out that use in favor of an equally effective, nontoxic alternative. That market shift reduced emissions of cancer-causing formaldehyde from batt insulation factories by about 90%—from about 600,000 pounds in 2005 to about 60,000 pounds in 2014 (the remaining emissions came from a small quantity of insulations sold to the industrial building market that continue to use formaldehyde).[6]

Sometimes companies cling to toxic chemicals because it's difficult to find effective alternatives. Chemicals that stick around for hundreds of years are very good at standing up to the elements, which is critical. "There are very high expectations for the performance of building materials," Bill said. "They have to stand up to heat, cold, moisture, contraction, expansion. No one wants to be the first to try some hippie-dippie roofing product, for example, if we don't yet know if it will hold up after fifteen years. And the stakes are very high—if a roof you built fails, that's a catastrophic failure."

The manufacturing industry runs on relatively small margins, so even when nontoxic alternatives emerge that can rival a toxic product in terms of performance, they often have a higher price point that manufacturers don't want to shoulder. "That's where political interventions become effective," Bill said. "That can be a ban, but it can also be as simple as penalizing the use of toxic chemicals by enforcing existing environmental laws regulating emissions and waste. If companies aren't able to externalize those costs to the rest of us, it costs more to create and dispose of toxic materials than it costs to use nontoxic ones."

In the mid-2000s, Bill and his colleagues at HBN realized they also needed to create public pressure for change by informing architects, designers, and building owners about what chemicals were being used in building materials and the health risks associated with them. Unlike food and cosmetics, building materials don't always come with a list of ingredients. They decided to launch a database that would allow people in the green building industry to look up a product and learn about any health hazards associated with it, but that proved difficult. There were multiple chemical databases out there, but most didn't connect to one another and most weren't available to the public. Some charged up to $30,000 a year for access. But Bill was determined. He hired Lawrence Kilroy, an old friend from Greenpeace, to help him tackle the problem. "I had no idea how we were going to do this and I thought it might be impossible," Bill said. "I knew I needed someone who would not be easily deterred. Someone who had that never-say-die energy you'd want in the pursuit of an IPO at a startup. Larry had that."

The project was an uphill battle, but the group's doggedness eventually paid off, and the database launched in 2006. It was dubbed Pharos, after the Pharos Lighthouse of Alexandria, one of the seven wonders of the ancient world and a symbol of light in darkness. Pharos taps into more than seventy data sources to provide tools that allow architects, contractors, and builders to look up the hazards of various building materials, compare the functionality of different materials, assess trade-offs, and collaborate with others. In 2009, Pharos earned the US EPA's Environmental Award for Outstanding Achievement, and in 2010 Google collaborated with HBN and used Pharos to develop a custom tool its architects could use while designing its offices around the world.

Pharos was also the first green building materials ranking system to incorporate environmental justice criteria. Its engineers did this by looking at health effects outside the building itself. Even if a product was perfectly safe for building occupants, if toxic chemicals were emitted during manufacturing or could leach into groundwater after disposal, it would not be rated as "healthy." The group received some pushback over this feature, so when HBN was invited to present at industry conferences, they would bring in speakers who lived near manufacturing plants. The speakers would share stories about what these plants had done to their communities—increasing rates of cancer, heart disease, COPD (chronic obstructive pulmonary disease), and asthma—and implore the architects to consider the ethics of marketing the products manufactured there as "green" or "healthy." It was an effective way to get architects on board with the rating system, who in turn pressured building material companies. "Through that strategy, we found a great number of allies and really influential leaders in the green building movement that have lasted a whole generation now," Bill said.

Those connections led HBN to collaborations with other groups pushing to incorporate environmental justice criteria in other green building standards, including LEED, which is now the most widely used green building rating system globally. They also worked with the International Living Future Institute to create the first "red list" of chemicals and materials that cannot be used for the Living Building Challenge, one of the most rigorous green building certification programs in the world. Certified Living Buildings aim not only to avoid negative impacts but also to "create a positive impact on the human and natural systems that interact with them."

HBN is also pushing for greater transparency in the building materials industry. Product formulations may change from year to year, and the toxicity of the same product can vary tremendously based on who manufactures it and where. One facility might make the same brand's product differently than another facility does.

"When we started there was industry resistance to disclosing ingredients," Bill said. "We've been chipping away at that, and it's beginning to change. We're trying to communicate that increasing transparency is in the companies' best interest because consumers are increasingly demanding it."

In the meantime, the team at HBN have become experts in the way materials like particleboard and linoleum are made. They may not always be able to say exactly what's in a specific brand's product, but they can provide consumers with a general sense of what is typically found in these types of materials.

"Transparency is the future for this industry," Bill said. "If you're not a leader in transparency in this industry, you're already playing catch-up. A lack of transparency can cost companies not only a loss of consumer confidence but also block their access to the best talent. Young people especially don't want to associate themselves with brands that end up on the front page of the *New York Times* because of a public health scandal."

————

After our tour of the Breene family home in Oil City, Berry drove me about 15 miles north to Titusville, where one of the world's first modern oil wells was drilled, to show me two of her other murals. On the way into town, we passed a small monument to Sun Oil Company—which later became the portmanteau "Sunoco"—complete with antique gas pumps and a restored antique tow truck painted in the company's trademark yellow and blue.

Berry's first mural was painted across several panels inset in brick archways on the side of a pub. The panels move through history from left to right. The first depicts Native Americans lighting the oil-slicked top of Oil Creek on fire—a ritual Berry said is part of local lore (though no one seems to know when or why it occurred, and I was unable to find a record of it having happened). A bald eagle soars overhead above an oil surveyor peering through a viewfinder and a man steering oil barrels down a bend in the river. The second panel pictures Colonel Edmund

Drake, the surveyor who discovered the site of the original well, sitting beside a spurting oil derrick. The third panel depicts the former Cyclops Steel plant in Titusville (the steel industry dominated the region for a period after the oil dried up); several historic buildings that no longer exist; and a portrait of Ida Tarbell, a local pioneer of investigative journalism whose work led to the breaking up of Standard Oil Company's monopoly. The final panel pictures a modern-day family kayaking on Oil Creek in orange life jackets. Further downstream there's a fly fisher, football player John Heisman (a Titusville native) in his signature trophy stance, and a little girl hula-hooping on the shore—an ode to the nation's first hula hoop, which was forged from petroleum-derived plastics in Titusville in 1958.

Looking at the hula hoop I thought about how today, plastics have become a lifeline for the fossil fuel industry. Most plastic is now made from chemicals sourced from fossil fuels. To make plastic, fossil fuels like coal and natural gas are heated to high enough temperatures to crack them into molecules that become the building blocks of plastic. For example, propane is broken down into propylene, which is used to make plastic bottles. A similar process is used to turn ethane into ethylene, which is used to create the polyethylene pellets used to make plastic bags and packaging. Despite growing recognition of the plastic pollution crisis, plastics manufacturing in the first thirteen years of the twenty-first century surpassed total production in the last century, and production is expected to double again in twenty years and almost triple by 2050. This is being driven in large part by the US fracking boom, which has prompted fossil fuel companies to build massive petrochemical complexes in order to sustain demand for natural gas, which is otherwise perpetually in oversupply. If that plan is successful, this cycle will create new markets for plastic and keep fossil fuel extraction and plastics manufacturing—both major sources of carcinogens—booming for decades to come.

A 2021 study by researchers in Switzerland identified 10,547 chemicals used in plastics manufacturing and found that about a quarter of them are "substances of potential concern"—meaning that they're carcinogens or endocrine disruptors, they can cause reproductive harm, they can damage specific organs or accumulate in the bodies of humans and animals, or they're toxic to marine life.[7] More than half (53%) of the chemicals the researchers identified as potentially concerning are unregulated in the United States, the European Union, Japan, and South Korea, and 901 of those chemicals are approved for contact with food in the United States, the European Union, and Japan. At least 1,000 of the chemicals the researchers tagged, including carcinogens, are harmful even in very small doses. And an additional 39% (4,100) of the chemicals used in plastics manufacturing haven't been studied enough to determine whether or not they're hazardous.

While Berry and I stood on the sidewalk outside the pub gazing up at the shiny red, plastic hula hoop in her mural, a woman who'd stopped at the intersection put down her window, observed that we must be from out of town, and proceeded to tell us where else we should go around town to see other murals and historic sites.

"Okay, thanks," Berry said, waving as she drove away before turning to me to shrug and say, "Small towns, you know?"

———

After tackling worst-in-class materials and industry transparency, the third phase of HBN's efforts has been aimed at getting toxic chemicals out of affordable housing. Affordable housing units are often in zip codes that are affected by pollution from traffic and industrial sources, and exposure to toxic building materials is the last thing these communities need.

Bill approached Enterprise Community Partners, a national nonprofit working on affordable housing, about a possible partnership. The organization had already begun following green building guidelines, so it was a natural fit. Enterprise updated its Green Communities Criteria, a green building certification for public housing products, using many of HBN's suggestions, which consider the toxicity of building materials not only for building inhabitants but also across their entire life cycle, from manufacturing to disposal.

Following the success of that program, HBN secured additional funding for work related to affordable housing and built a consortium with other groups, including the New York Parsons School of Design, the Health Product Declaration Collaborative, and the Green Science Policy Institute. In 2016 HBN launched the HomeFree project, with demonstrations in Seattle, Minnesota, New York, California, and Washington, D.C. The projects sourced nontoxic, regionally appropriate building materials with the goal of reducing affordable housing residents' likelihood of experiencing asthma, developmental delays, and cancer. Leaders at each site shared insights with other local groups doing affordable housing projects. After those pilot projects, HBN formed additional partnerships with more groups, including the Housing Partnership Network, Stewards of Affordable Housing for the Future, and the Natural Resources Defense Council's Energy Efficiency for All program. The goal was amplifying regional efforts to advance environmental justice at the national level.

Historically, the green building movement has had a trickle-down mentality: organizers would try to get companies with big budgets, like Google, to pioneer and use healthy building materials with the hope that eventually there would be less toxic manufacturing in poor neighborhoods, and the market for those materials would expand so prices could come down for everyone else.

"Our affordable housing programs provided an opportunity to leapfrog over that trickle-down theory," Bill said. "Now we're directly improving the health of people of color and lower incomes and people living in oppressed, toxic zip codes."

In 2022, Bill left HBN to become the executive director of the Passport Foundation, a California-based grantmaking institution that funds environmental and social causes, and a long-time supporter of HBN. His departure came after a multiyear transition to new executive leadership, and HBN is still going strong in his absence. Bill's new position enables him to stay in the loop about what's going on at the organization, and help fund similar organizations that are working toward systemic change.

Around the same time, Bill's former colleagues began a process to expand the use of Pharos and the robust data ecosystem that they'd developed to provide increased transparency about chemical safety in products used not only in buildings but also in other sectors, like food packaging and personal care products. HBN started the process of rebranding to reflect expanding it's focus beyond just healthy building materials—a process that will eventually include a new name for the organization.

Before any of those changes, in the spring of 2021, Bill and his wife had recently moved from Vermont to Long Beach, California, to be closer to their three-year-old granddaughter. Bill gave me a remote tour of the home they'd been renting and eventually purchased. They were still moving in, and there were unpacked boxes and extra pieces of furniture strewn across sunny rooms with cream-colored walls. Plants of various sizes brightened many corners of the house, and palm leaves shimmered in the breeze outside the windows. The mantles, bookshelves, hutches, and desktops were already full of framed family photos, artwork, and mementos, giving the place a cozy, lived-in feel.

The one-story house is beige stucco, which Bill was happy to see in lieu of vinyl siding because it's less toxic in terms of production, occupant exposure, and disposal, and is breathable in hot weather. Most of the house had hardwood floors, but in a bathroom littered with his granddaughter's brightly colored bath toys in the shapes of fish, ponies and boats, Bill pointed out that while the floor looked like tile, it was actually one large sheet of vinyl.

"It's old enough that it must contain a lot of phthalates," he said. "It's not a very big surface area so we may look at replacing it with real tile. The dilemma is that it's still very serviceable and could probably last another fifty years. My granddaughter is only in there occasionally for a bath, not generally spending a lot of time on the floor, so I'm not overly concerned about exposure, and ripping it out kind of feels like a waste—especially knowing there's no real safe way to dispose of it. It's one of those dilemmas I know too much about to be happy with either choice."

There are a lot of those types of dilemmas when it comes to trying to make your own home less toxic. Another example is wall-to-wall carpeting. It's unhealthy because it off-gasses chemicals and traps air pollutants and bacteria. But ripping it out poses its own threats— when old carpet is pulled up, the history of a home's toxic exposures is unearthed in the resultant dust, which can expose workers or family members to more of those compounds than if the carpet had stayed in place. While occupants would still likely benefit from taking carpets out in the long-term, contractors rarely wear adequate protective gear while ripping it out, and there's no environmentally friendly way to dispose of old carpeting.

"As a family we do the best we can to be educated and reduce our exposures, but I don't think of myself as living an exemplary life when it comes to toxics," Bill said. "That's mostly for financial reasons—my wife and I have both worked in the nonprofit world all our lives, so it comes down to a decision between creating a retirement account or living an ultra-organic life."

"It's that way for most people," he added. "The vast majority of people would prefer to avoid toxic chemicals to the maximum extent possible, but the barriers to entry can be substantial."

It's easier than it used to be to find healthy building materials, but it can also be hard to tell which products claiming to be "green" are legitimate. It takes a lot of time and research, which most people don't have the capacity to do for every little decision about grout, carpeting, and window sealant.

"It's just not possible for every person who is concerned about this to do this kind of deep-dive research before undertaking a home or office renovation project," Bill said. "It's one of the main reasons I work so much on driving systemwide change."

CHAPTER 5

B. Braun: Safer Medical Treatment through Innovation

On a bright, frigid day in January 2022, I stood in the hallway of a medical supply factory in Allentown, Pennsylvania, peering through a window at a worker in a full-body cleanroom suit and safety glasses. She was bent over a table fidgeting with long strips of plastic tubing, occasionally glancing at the detailed work order propped in front of her. She looked up and we exchanged polite waves before she returned to her task, which involved applying adhesive to the tubes and attaching tiny plastic connectors to each end.

Mounted on the wall beside me were acrylic bins labeled "hairnets," "beard covers," and "wear your ear plugs." A rack stashed sneakers and boots for workers who had switched to company-issued, cleanroom footwear. Across the hall, a vinyl door covering loudly retracted upward, revealing a warehouse of shelves filled with plastic bins. Small forklifts maneuvered back and forth, following the most-efficient routes for product retrieval that an algorithm had programmed for them. Through another window, I watched a conveyor belt carry tiny pieces of silicone tubing around in a circle. A robotic arm at the start of the line punched out small circular filters, and as the tubes revolved, another robot arm precisely placed a filter into each piece of tubing. At the next stop, a third arm placed a small washer on top of the filter, then the tiny tubes were whisked along to a fourth arm that ultrasonically welded the filter into place, before a final arm plucked each one up and dropped it into a shoot for transport to another part of the assembly line. The whole procedure happened in the span of a few seconds.

"Production is actually very slow right now," John Grimm, vice president of research and development at B. Braun, told me. John has thinning gray hair, blue eyes, and rosy cheeks. He wore a fleece jacket over a button-down shirt with slacks and brown tasseled dress shoes and cheerfully greeted just about everyone we passed in the hallways of the factory.

"A lot of this is automated but as you can see, a lot of it still requires workers," he said. "There are full lines we can't run because so many people are out sick with omicron. It's a real struggle."

The omicron variant of COVID-19 was surging throughout the country, and to enter the building I'd had to complete a questionnaire about potential exposures and symptoms, stare into a camera for an ID sticker I was instructed to wear for the duration of my stay, and don a blue surgical mask on top of the N95 mask I was already wearing. I knew John and some of his colleagues had been mainly working from home since 2020, and I expressed surprise at the crowded halls. "It's a factory," he said, "so most workers have to be here in person. And it's essential."

It's essential because B. Braun manufactures everything from IV bags and tubing to epidural and anesthesia trays, syringes, and pharmaceutical products. At this location, 5 miles from the company's US corporate headquarters in Bethlehem, Pennsylvania, they also make lots of connectors—the small plastic and silicone ports, valves, and clamps used to connect IV bags to patients in hospitals around the world. B. Braun's customer base is broad. It includes hospital systems, manufacturers of

medical devices, pharmaceutical companies, startups, veterinary product companies, and group purchasing organizations (GPOs), which buy
supplies for networks of hospitals to get discounts based on collective
buying power. "We make about 300 of these per minute here," John said,
holding up a small, pale blue connector port. The piece is no bigger than
a blueberry, but its production keeps the US medical system running.

Bethlehem is a small city of about 75,600 people along the Lehigh
River in eastern Pennsylvania. In the early 1900s, Bethlehem Steel was
the second-largest steel company in the country. Now defunct, its hulk-
ing, rusty furnaces still tower over the river, visible from many places
in town. Coal extraction also played an important role in the region's
history; one Allentown high school's mascot is the Canary. Today Bethlehem is home to a couple universities, and its main street is full of
breweries, restaurants, and boutiques amid historic landmarks. On the
outskirts of town are several industrial manufacturing parks, including
B. Braun's, where around 1,150 people work in the factory, and about
150 people, including John, work in the adjoining corporate offices. B.
Braun–affiliated companies also have manufacturing plants in California, Texas, and the Dominican Republic.

In addition to the factory floors, the plant is home to various laboratories, where engineers, scientists, and technicians can step away from
their cubicles, don safety glasses, and test new products. They might test
products' durability with an "aging chamber" that uses higher temperatures to speed up the natural aging process, or use specialized equipment
to measure the tensile strength of IV and catheter tubing, or run assays
to determine the BPA content of a piece of sample material. Outside the
manufacturing floors are display containers filled with tiny white, blue,
and clear pellets of plastic resin that will eventually become valves, ports,
clamps, and other products.

Every day, plastic products like the ones manufactured by B. Braun
help save people's lives. Plastics are used in everything from syringes and

stents to hearing aids and prosthetics. They're used to create IV bags that deliver chemotherapy drugs to cancer patients and tubing that delivers oxygen to premature babies.

They are also, in many cases, made with potentially cancer-causing chemicals that can make their way into patients' bodies during treatment. Phthalates like di(2-ethylhexyl) phthalate (DEHP), which is commonly used as a plasticizer in polyvinyl chloride (PVC) IV bags, tubing, and medical devices, can disrupt hormones, interfere with reproduction, and raise cancer risk, and patients can receive high doses of these compounds when receiving medical treatment.

A 2021 study found that among patients being treated for breast cancer, those with higher levels of DEHP metabolites were more likely to relapse following treatment, and patients with a certain genetic marker who had higher levels of DEPH metabolites were also more likely to die.[1]

A 2020 assessment of sixteen years of studies on endocrine-disrupting chemicals in medical supplies and medications summarized evidence that during treatment, patients can receive concerningly high doses of these chemicals.[2] Endocrine-disrupting compounds have been linked to infertility, heart disease, stroke, immune system dysfunction, and cancer, but given the complex, and oftentimes delayed, impacts of these chemicals on health, few studies have been able to directly connect these exposures to health outcomes. The authors concluded that failure to disclose these risks to patients violates core medical ethics, but that most physicians aren't aware of them. They also emphasized that the risks from these medical exposures are likely understated because our current knowledge is restricted to only a few categories of chemicals and medical devices.

Most health care providers aren't trained in talking to their patients about the link between toxic exposures and cancer.[3] And those same nurses and doctors are exposed to endocrine-disrupting and carcinogenic chemicals every day in the course of their work. Many of the cleaning products and disinfectants used to keep hospitals and doctors' offices sterile contain carcinogens and endocrine disruptors, such as formaldehyde and phthalates. Flame retardants containing endocrine-disrupting chemicals are still commonly used on furniture, bedding, and curtains used in medical facilities, and medical devices, surgical tools, and medications can also contain chemicals that raise cancer risk. It's surprising that places devoted to healing are so full of potentially harmful chemicals.

John and others at B. Braun recognize the irony. The company is working to replace harmful chemicals in medical supplies like IV bags and catheters with safer alternatives. I got in touch with John through my employer's parent organization, Environmental Health Sciences, which has, through a consultant, helped advise B. Braun on their plan.

B. Braun was the first IV fluids manufacturer to remove DEHP from most of its products in the 1990s, and as of this writing it remains the only supplier that offers a full line of IV solution containers that are not made with PVC or DEHP. In recent years, B. Braun has overhauled its material sourcing procedures to ensure that they know exactly what's in all of their products, and the company hopes to eventually replace potentially harmful chemicals in everything they make.

B. Braun was founded in 1839 when Julius Wilhelm Braun bought the Rose Pharmacy in Melsungen, Germany, and began selling medical herbs by mail. Following the success of that venture, the Braun family built a manufacturing plant and started producing surgical sutures and tools, which they sold to hospitals. The business expanded rapidly, and B. Braun created some of the first industrially manufactured IV solutions, developing the first disposable plastic container for IV solutions in 1956. In 1971 the company produced the first automatic infusion pump capable of regulating the amount of fluid entering a patient from an IV bag. Today, the company, in which the Braun family still has an ownership interest, is among the largest medical supply manufacturers in the world, with subsidiaries on every continent. B. Braun's journey toward creating healthier medical supplies started, like so many technological advances do, with a happy accident.

In the 1990s, a California-based medical supply company called McGaw, Inc., set out to make a clear IV bag. At the time, IV bags were mostly opaque, making it difficult to see how full they were, and McGaw saw an opportunity. The company successfully engineered the first glass-clear IV bags by eliminating the use of PVC and DEHP, since both compounds gave the bags opacity.

In 1997, B. Braun acquired McGaw for around $320 million. By then, emerging research had begun to link DEHP and PVC to increased cancer risk, and B. Braun saw a new advantage. "We realized we had an opportunity to promote our products as safer, more sustainable, and more favorable to the patient," John said. "That wasn't the original goal when McGaw was developing these IV bags, but it stuck out to us as a huge benefit."

B. Braun later launched IV sets made without DEHP or PVC. IV sets include all the parts that carry fluid from an IV bag into a patient—needles, tubing, and catheters. This was challenging because it's incredibly complex to engineer an IV bag and all its parts. You must ensure that microbes can't enter the bag, that there's no flaking of the plastic into the fluids inside the bag, that bags that are dropped or shaken don't burst or leak, and that any chemicals used in connectors, ports, or printed and adhesive labels won't have dangerous interactions with chemicals in the IV bags. "Everything that goes into that package needs to be well understood," John said. "We have to consider every possible way that a chemical or contaminant could possibly migrate and contaminate the product."

Sometimes, he said, during leachability studies, they find impurities formed from combinations of various chemical components that they can't even readily identify. When that happens, they head to the lab to figure out what the chemical is, where it's coming from, and whether it's harmful (and whether they have to go back to the drawing board). "A lot of work goes into piecing together the molecular structure of something you can't identify on the periodic table of elements," John said.

John comes from a family of nurses: his sister is a nurse, as was his mom, his grandmother, his great-grandmother, and his great-great grandmother. He studied industrial engineering in the 1980s and got a job as a plastic injection mold designer shortly after graduating, helping create parts for everything from consumer goods and medical supplies to the telecommunications and automotive industries. "I noticed I always seemed to be more enthusiastic about designing something for the medical field," he said. "Those assignments came with a certain elevated sense of importance."

A few years later he took a job at B. Braun, where he's been working for the last thirty-five years. One of the earliest projects he worked on in the 1990s was the first needle-free technology for IVs, which helped prevent accidental sticks of health care workers that increased their risk of blood-borne diseases like HIV and hepatitis. He was proud of that project, knowing he was helping to keep people like his mom and his sister safer on the job. "I take pride in being in the health care industry," he said. "It's humbling to know that you're contributing in a small way that becomes part of a bigger picture that's all about helping people."

———

Berry showed me a second mural she had painted in Titusville, this one on the side of a rusty railway bridge along the Queen City Trail. "The Queen City" was one of Titusville's nicknames during its peak, back when it boasted mansions and amenities that according to local lore, locals thought were fit for a queen.

This mural struck me as being visually different from her other murals in the region—the color palette is more muted, all pale greens and blues—but there were common themes. This one, too, features a river, a kayaker, and oil derricks spurting teardrops of oil. The mural also includes a bear, a blue heron, kids riding bikes, a leaping trout, and lots of white and pink star-shaped mountain laurel, the state flower. While we looked at Berry's work, a train rumbled by and walkers and bikers cruised along the trail. It might seem odd to portray oil derricks surrounded by flowers and wildlife, but it's the way things are here: oil is considered as much a part of the region's natural bounty as birds, trees, rivers, and plants.

On our way back to Oil City, Berry and I stopped by the Drake Well Museum. After the famous oil well dried up in 1861, the site was abandoned, and nothing was left behind but some rusty pieces of pipe. The property changed hands a few times, then in 1914 a local chapter of the Daughters of the American Revolution erected a sandstone boulder with a bronze plaque commemorating the site. That commemorative boulder is still there today, along with a small museum, a gift shop, and a park full of wooden replicas of oil derricks and engine houses, a grave-yard of old oil pumping machinery, and remnants of the world's first oil pipeline, all nestled in a pretty, wooded valley beside Oil Creek, with meandering pathways that connect to the Queen City Trail. A plaque installed by the American Petroleum Institute in 1959 commemorates a man named Samuel Van Syckel for building the world's first oil pipe-line nearby. After praising his technological achievement, it concludes, somewhat ominously, "Thus was set in motion a revolution in transpor-tation whose ultimate results are not foreseeable even today."

To enter the museum or the park you must pass through the gift shop, which sells keychains, hats, T-shirts, magnets, and mugs commemorat-ing the "world's first oil well," thick tomes on local history, miniature oil derricks and gas well pumpjacks in the form of toys and puzzles, and small glass bottles full of black liquid labeled "Pure Pennsylvania Crude Oil." Visiting with Berry, her hair still short from chemo, and understanding how the oil industry has ravaged the environment, climate, and human health, I couldn't help but experience the museum as a bit grisly—akin to a war museum absent any mention of violence—despite a few recent additions to the exhibits that vaguely nod toward "a sustainable energy future."

An exhibit titled, "There's a Drop of Oil and Gas in Your Life Every Day!" celebrated the presence of petroleum-derived products in everything from cosmetics to cars, clothing, and plastics, boasting that "abundant oil and natural gas have made our lives easier," and calling to mind the scale of the global plastic crisis. Consumers throughout the world have been led to believe that as long as they put their oil-derived plastic waste into a recycling bin it will be given new life. In reality, only 10% of the plastic produced on the planet has ever been recycled.[4]

It just isn't cost-effective to recycle plastic. It takes time and energy to sort, and plastic degrades during the recycling process and loses value. An investigative report published in 2020 by NPR and PBS *Frontline* found that industry insiders knew as early as the 1970s that recycling wasn't a meaningful solution to the growing problem of plastic waste.[5] One executive even wrote in a 1974 speech, "There is serious doubt that [recycling plastic] can ever be made viable on an economic basis." Despite that understanding, the industry used the promise of recycling to sell billions of dollars of new plastic, all while telling consumers that responsibility for keeping plastic out of the environment was theirs alone. Plastic that isn't recycled ends up in landfills, oceans, or incinerators, resulting in carcinogens in plastics leaching into soil and groundwater and finding their way into fish and air pollution—all of which increases people's exposure to these chemicals. Even more troubling, plastic that *is* recycled can become more toxic if recyclers don't take care to separate different types of plastics—something most recyclers currently lack the capability to do. And plastics are often manufactured, recycled, and disposed of in low-income neighborhoods and communities of color, adding to the disproportionately high levels of exposure to carcinogenic pollutants that these communities already face.

The friendly, middle-aged woman working at the museum gift shop recounted a tour guide telling a little girl that everything she was wearing—from her hair bow, to her dress, to her shoes—was made with products derived from petroleum, which she said, grinning enthusiastically, made the girl "wide-eyed, just totally shocked."

————

On March 6, 2020, something happened that called into question B. Braun's commitment to environmental health: lawyers filed a class action lawsuit alleging that emissions of a chemical known as ethylene oxide from the Allentown plant had caused more than a dozen cancer cases and three cancer-related deaths among people who worked at the facility or lived nearby.

Local news articles profiled heartbroken cancer patients, attorneys promising to see justice served, and residents who were newly fearful for their health after learning about the lawsuit. One article featured a plaintiff who, having been diagnosed with leukemia in 2018 after working at B. Braun for eight years, had run out of treatment options.[6] He said he cried every day thinking about his young daughter growing up without her father.

B. Braun denied the allegations, saying there was no evidence that its emissions have caused cancer. The company is defending itself against the lawsuit, which was still pending in the fall of 2022. Company spokespeople have also maintained that B. Braun's use of ethylene oxide is required to deliver life-sustaining medical products, and that they use it responsibly and in compliance with all regulations.

Ethylene oxide, a gas that's derived from petroleum or natural gas, is used to manufacture plastics, detergents, adhesives, and pharmaceuticals, among other things. It's often emitted from petrochemical and chemical manufacturing facilities, and it's used to sterilize herbs and spices in the US (the EU has banned its use on food products). The FDA also requires the use of ethylene oxide to sterilize certain types of medical equipment. Ethylene oxide is used on about 50% of all US medical equipment that requires sterilization—including most of what is manufactured at B. Braun's Allentown plant.

In May 2022, the Pennsylvania Department of Health published an analysis of cancer rates near B. Braun's Allentown facility that did not find patterns typically associated with environmental exposures but noted that due to limitations related to the state's cancer registry data, the agency "cannot determine whether cancer incidences near the B. Braun facility were caused by [ethylene oxide] exposure." The analysis also stated that in the absence of real-time air monitoring data, the agency "cannot determine the levels of [ethylene oxide] that people inhaled or are inhaling and the associated human health risks."

But the problem of ethylene oxide is much larger than what's happening in Pennsylvania, which mirrors what has happened at medical sterilization plants across the country. It's a profound case of regulatory failure, one in a long line of failures by the United States to protect its citizens from cancer-causing chemicals.

Studies dating back as far as the 1970s indicate that ethylene oxide might cause cancer, and anecdotally, people have long understood its risks. In the 1970s and 1980s, hospitals noticed that employees who used ethylene oxide to sterilize equipment were getting sick and started sealing off the rooms where it was used, according to a toxicologist interviewed about the issue in 2019.[7] The science linking ethylene oxide exposure to cancer, particularly lymphoma and leukemia, became even clearer in subsequent decades.

In 1994 the World Health Organization gave ethylene oxide its highest cancer risk classification, naming it "carcinogenic to humans." It took the US Department of Health and Human Services another six years to upgrade its own classification of ethylene oxide from "reason-ably anticipated to be a human carcinogen" to "known to be a human carcinogen" in 2000. It wasn't until 2016 that the EPA added ethylene oxide to its Integrated Risk Information System (often referred to as IRIS). When they did, the agency determined that ethylene oxide was thirty times more carcinogenic than it had previously recognized. But it took two more years after that for the agency to publish an updated review of Americans' exposures to ethylene oxide through the 2018 National Air Toxics Assessment. That report looked at emissions from 2014 because it typically takes the agency several years to validate and review such a high volume of data from polluting facilities across the nation. So four years after the exposures had already occurred, the EPA identified twenty-five communities as having cancer risk from ethylene oxide exposure that exceeded 100 cancer cases per one million people, including Allentown, Pennsylvania, near the B. Braun plant.

Ethylene oxide use and emissions were already regulated by both the Occupational Safety and Health Administration and the EPA's National Emission Standards for Hazardous Air Pollutants, but finding cancer risk greater than 100 in a million from a single toxic chemical represents an unacceptably high risk according to the EPA, a benchmark that allows the agency to quickly take new steps to protect public health. Four years later, though, the agency had yet to enact comprehensive new regulations on ethylene oxide to address those risks. This kind of delay in regulating potent cancer-causing chemicals is the norm in the US.

"The air toxics office has seen long-standing delays in fulfilling its legal obligations across multiple administrations," said Emma Cheuse, an attorney for the environmental legal advocacy group Earthjustice, who has worked on numerous lawsuits aimed at compelling the EPA to revise its regulations on ethylene oxide. "That has happened in part because it's been historically underfunded and underresourced, and in part because of a lack of political will."

Throughout all these regulatory delays, some of the industries and trade groups that use ethylene oxide have challenged the EPA's analysis of the chemical, attacking the agency's review of the scientific literature on ethylene oxide and cancer risk and commissioning their own studies aimed at proving that the chemical is safe. The agency has had to spend time and resources defending its assessment, which has resulted in additional delays.

All these factors may have contributed to the EPA's inconsistent follow-up with the twenty-five communities it identified as being at high risk from ethylene oxide in the 2018 National Air Toxics Assessment, which presents a clear example of environmental injustice perpetrated at the federal level. A 2021 investigation by *The Intercept* found that of the communities living near the twenty-five "high-priority facilities" identified by the EPA in 2018, those with populations that were at least 60% white and had an average per capita income higher than $30,000 per year were nearly three times more likely to have been informed about their risk from ethylene oxide than communities that were less than 60% white and had an average per capita income under $30,000.[8]

One example highlighted in that investigation stands out. After the report was released, lawyers and activists in Willowbrook, Illinois, a wealthy Chicago suburb, launched a campaign against a local medical sterilization facility that was emitting ethylene oxide. Their campaign was so successful that the state's newly elected governor took up the cause, taking advantage of a rarely used executive authority to temporarily shut down the plant. The then head of the EPA, Andrew Wheeler, met with state officials, and Bill Wehrum, the chief of the EPA's Air and Radiation Office, traveled to Willowbrook for a community meeting with concerned residents. New state laws strictly regulating ethylene oxide emissions were passed quickly with bipartisan support. It was a resounding success story for the residents of Willowbrook.

Residents of St. John the Baptist, a small, mostly Black community in Louisiana, were not as fortunate. Some census tracts faced a cancer risk of 317 cases per million people from ethylene oxide, according to the EPA's national air toxics report—significantly higher than the census tract with the highest cancer risk from the chemical in Willow-brook, which was 251 per million. And St. John the Baptist also had problems with many other cancer-causing chemicals from industrial pollution. Its most at-risk census tract in the national air toxics report had a total cancer risk of 1,505 cancer cases per million people, the highest in the nation. For comparison, the most at-risk census tract in Willowbrook had a total cancer risk of 282 per million. But it took years for anyone from the EPA to meet with the community in St. John the Baptist. No plants were shut down. The governor did not get involved. No new state laws regulating the chemical were passed. Four years after learning about their exposure, the community in St. John was still in more or less the same position as when the EPA's report first came out.

The risk was also higher outside the B. Braun plant in working-class Allentown, Pennsylvania, than it had been in Willowbrook. The most at-risk census tract near the Bethlehem plant had a total cancer risk of 596 per million in the EPA's 2018 report, with nearly all of it stemming from the plant's ethylene oxide emissions. But four years later, the EPA had still not held a public meeting with residents near the plant, and little had happened at the state level to address the problem.

––––––

After our tours through Oil City and Titusville, Berry took me to Pithole, another former oil boomtown nearby. Pithole was home to the world's first oil pipeline, but that's not what it's famous for today. It's known for being a ghost town that vanished as quickly as it had emerged, going from being a small city with a population of around 20,000 in 1865 to having a population of zero in the span of a few decades, according to news reports at the time.

Today, almost nothing remains of the original Pithole City. The site is hard to find, in the middle of acres of woods down a series of winding back roads. It's in a clearing marked by a modern-looking visitor's center that stayed closed throughout the pandemic. Thanks to some historic preservation initiatives, the grounds are well kept. Watching the tail end of a wintry pink and gold sunset, Berry and I wandered the mowed paths cut through tall grass representing former roads, reading the wooden signposts bearing the old street names and plaques marking the sites of former banks, hotels, and post offices.

Ghost towns like Pithole are rare, but western Pennsylvania is full of other towns that were once home to booming industries—coal, oil, steel, natural gas—but now face economic hardship and the lingering effects of pollution. Economists use the phrase "the resource curse" or "the paradox of plenty" to refer to nations that are wealthy in natural resources but fail to reap economic and social benefits from it. In recent years, some writers and researchers have also begun using the term to apply to Appalachia. But despite these oft-repeated examples of extractive industries exploiting the region's natural wealth and leaving only poverty and illness in their wake, the cycle continues today.

In 2016, Royal Dutch Shell began construction on a massive ethane cracker in Beaver County, about 24 miles northwest of Pittsburgh and 90 miles southwest of Oil City. The plant, which was expected to begin operations in 2022, will eventually convert a massive volume of natural gas and liquids into 1.8 million tons of polyethylene every year, much of which will likely be used to make single-use plastic bags and plastic packaging. It's one of up to five such facilities that have been proposed in the Appalachian basin region spanning Pennsylvania, Ohio, and West Virginia (though as of this writing, all but the Shell project had been canceled or put on hold, which some financial experts have attributed to unfavorable market conditions).

The facility will demand natural gas liquids from an estimated 1,000 new fracking wells. Fracking also raises cancer concerns. A 2017 study by researchers at Yale and NIH found that fracking wells emit fifty-five chemicals that are known to or may cause cancer, including twenty that have been shown to increase the risk of leukemia and lymphoma.[9] Other research indicates that living near oil and gas wells increases the risk of childhood leukemia.[10] Pennsylvania is second only to Texas in natural gas production, and a rash of cases of a rare childhood cancer in four of western Pennsylvania's most heavily fracked counties has prompted numerous investigations in recent years.[11]

The petrochemical facility itself will also raise the cancer risk in the region. It's the same type of facility that earned Louisiana's "Cancer Alley" its moniker. Cancer Alley is an 85-mile-long industrial corridor between New Orleans and Baton Rouge that's home to more than 140 petrochemical plants and where cancer risk from air pollution can be as high as fifty times above the national average.[12] The region is home to St. John the Baptist, the community experiencing ethylene oxide exposure, and uncoincidentally, the highest cancer risk from air pollution in the nation, and many communities like it—poor, Black, and exposed to extraordinarily high levels of pollution.

When it was permitted, the Shell ethane cracker in Pennsylvania didn't anticipate generating ethylene oxide emissions, but that could change if the plant shifts its operations in response to market pressure down the line, and it will emit a slew of other cancer-causing and otherwise harmful chemicals, including benzene, toluene, hexane, formaldehyde, and ammonia. Local environmentalists, community advocates, and Indigenous groups spent a decade fighting to stop the Shell ethane cracker and the petrochemical buildout, but their efforts have been unsuccessful. Similar stories have been repeated in communities the world over.

Berry and I were the only people in Pithole on that December evening. As we walked through the barren field that had once been a city, the sun disappeared behind the leafless trees, and a bright half moon rose against the pale sky. Berry pointed out a steep hill she had once snowboarded down with friends as a teenager. As the sun faded, the air suddenly turned sharply cold, making Pithole feel even more ghostly, and we hurried back to the warmth of Berry's car.

————

When the EPA still hadn't notified many of the less affluent communities about their high cancer risk from ethylene oxide in March 2020, the agency's Office of the Inspector General published a public "management alert" calling on the agency to take prompt action in these communities. The notice got some media attention, but the EPA's response was still slow to materialize, exacerbated in part by the pandemic, according to Cheuse. During all these delays, class action suits like the one against B. Braun have cropped up in communities across the country.

"Cancer risk accrues over a long period of time, but these communities have already been exposed for decades," Cheuse said. "The idea that people who've experienced cancer or the death of a loved one should have to sue after the fact for a remedy that's never going to make them fully whole is absolutely not the framework we're supposed to have in the United States. It's an important legal right we're supposed to have to stop this kind of harm before it happens."

From one perspective, B. Braun has gone above and beyond what's required by law to minimize its ethylene oxide emissions. In 2018, when the EPA published its assessment of cancer risk from the chemical in different census tracts, B. Braun was the twelfth-largest emitter of ethylene oxide in the country. Between 2008 and 2018 the plant emitted an annual average of almost 5,000 pounds of ethylene oxide per year, peaking at 8,960 pounds in 2009. But all those emissions remained well below the 20,000 pounds it's always been legally allowed to emit under its permit from the Pennsylvania Department of Environmental Protection, and in 2020, B. Braun finished installing a new, best-in-class filtration system that destroys more than 99% of its ethylene oxide emissions, despite having no regulatory mandate to do so.

From another perspective, this was too little too late. The potential dangers of ethylene oxide have been known by the industry since the 1970s and formally acknowledged by public health agencies since the 1990s. The filtration technology B. Braun installed to capture its ethylene oxide emissions is expensive, but it has been available since at least the 1990s, according to industry experts. Both B. Braun and the Pennsylvania Department of Environmental Protection have touted the new filtration system as a solution to the plant's ethylene oxide emissions, but some experts say ethylene oxide is so carcinogenic that the system might not actually be enough to protect surrounding communities from cancer risk.

"Focusing on emissions reductions is a bit of a red herring," said Todd Cloud, an attorney and regulatory compliance expert who worked as a consultant for the fossil fuel and petrochemical industry for twenty years before shifting to consulting for environmental advocacy groups. "This stuff is extremely carcinogenic, so in many cases anything short of a 100% reduction is not enough. A 99.99% reduction in emissions sounds really good, but if you were to model the remaining concentration of ethylene oxide in a nearby neighborhood or school, it would likely still be creating an unacceptably high cancer risk under the EPA's standards."

B. Braun disagrees that this is the case with their facility, but Cloud also noted that these reductions mostly account for point source emissions, while fugitive emissions—unintentional leaks from places other than filtered pollution stacks—are a huge problem when it comes to ethylene oxide. Facilities have to report their best guesses about fugitive emissions to the EPA, but that process is "more voodoo than science," according to Cloud, and without continuous fence line air monitoring, communities near facilities that use any ethylene oxide at all are unlikely to have an accurate understanding of their cancer risk. It isn't possible to capture 100% of emissions in facilities where ethylene oxide is used.

These are some of the reasons that groups like Earthjustice are pushing the EPA to completely ban ethylene oxide for medical sterilization. The FDA has identified a few chemicals, including hydrogen peroxide, as potential replacements, but as of 2022 the agency still required the use of ethylene oxide. Even if the agency approves replacement chemicals, it could take a long time for the transition to be implemented.

"These facilities are in a tough spot," Cloud said. "They have one federal agency telling them they have to use this chemical and another telling them not to use it because of the cancer risks, so they don't entirely know what to do."

In public statements, B. Braun has defended its use of ethylene oxide and denied liability for the cancer cases included in the class action suit. The company also launched a website about its use of ethylene oxide that touted B. Braun's safety record and contributions to life-saving medicine, emphasized its recent reductions in ethylene oxide emissions, and implied that the chemical poses no risks to workers or the community.[13] The website cited and linked to an industry trade group page that contested the EPA's assessment of the cancer risks associated with ethylene oxide. B. Braun is a member of that trade group, Advanced Medical Technology Association, or AdvaMed, which is among the organizations that have worked to undermine the EPA's science on ethylene oxide's cancer risk.

Despite these attacks and tremendous pressure from multiple industries, the EPA has maintained through multiple presidential administrations that its assessment on the cancer risks associated with ethylene oxide is sound. "If they didn't renounce it under Trump, they're unlikely to back down now," Cloud said.

The department that's responsible for emissions of ethylene oxide at B. Braun is separate from the research and development group that's working to get harmful chemicals out of medical supplies. John Grimm is confident that the new filtration system is adequately protecting the surrounding community and workers, including himself and his teammates who work in the building. A new EPA analysis published in 2022 lowered the 2018 assessment of cancer risk to around 60 in 1 million in the area surrounding B. Braun's Allentown plant. And the ambitious work John's team is doing to eradicate cancer-causing chemicals from medical supplies is no less important in light of the issues surrounding ethylene oxide.

The disconnect between B. Braun's dedication to eradicating cancer-causing chemicals from its products and its legacy with ethylene oxide illustrates a fundamental truth about the US chemical industry and cancer risk: even when market pressure encourages companies to prevent harmful exposures, scientifically sound regulations are the only sure path to widespread cancer prevention. "We've seen again and again that regulations drive advances in technology," Cheuse said. "If you say it's a problem to fix, American businesses are very innovative in figuring it out. If you don't say it's a problem, it's easier to stay the course and just keep doing what they've always done."

"We don't want people getting cancer from an industry that's trying to serve public health," she added. "EPA needs to figure this out and do its job." Only state agencies or the EPA have the power to protect all Americans from harmful ethylene oxide emissions. But regulatory shifts outside the US have led to major developments in safer chemicals for medical supplies at companies like B. Braun.

For example, in May 2021 the EU enacted a suite of new medical device regulations, referred to collectively as EU MDR, which was originally passed in 2017. Medical devices include everything from stretchers and tongue depressors to IV devices, pacemakers, implants, prostheses, and stents, and the legislation has an impact on many aspects of manufacturing, labeling, and distribution of medical devices, including the use of chemicals that are carcinogenic, mutagenic (capable of inducing genetic mutation), endocrine disrupting, or known to cause reproductive harm.

EU MDR forces the US medical supply industry to become safer because most big companies, including B. Braun, need to be able to sell their products in both the US and the EU. This pressure is strengthened by the recent implementation of numerous other EU regulations that limit the use of cancer-causing chemicals. Examples include REACH (Registration, Evaluation, Authorisation and Restriction of Chemicals), which applies to all industries in the EU; RoHS (Restriction of Hazardous Substances Directive), which applies specifically to electronic devices sold in the EU, including medical devices like pumps and dialysis machines; and MEAT (Most Economically Advantageous Tender), which asks major purchasers of products and raw materials to prioritize things like quality, social value, and environmental sustainability when buying materials instead of always choosing the cheapest option.

The US does not have federal equivalents of these regulations, but some state laws work similarly, like California's Proposition 65, which was adopted by California voters in 1986. The law, officially called the "Safe Drinking Water and Toxic Enforcement Act," required state health agencies to compile a list of substances that could cause cancer or reproductive harm, and required businesses to provide warning labels when those chemicals are present at high enough levels to pose a risk in products sold in the state. Most Americans have seen the familiar warning labels that read, "WARNING! This product can expose you to chemicals which are known to the State of California to cause cancer and birth defects or other reproductive harm." The list is updated once a year and now includes approximately 900 chemicals. The law regulates chemicals across industries and is more stringent than any federal-level legislation of toxic chemicals in the US, so it often effectively protects residents of other states too.

In the absence of meaningful federal regulations, B. Braun leads the industry when it comes to removing harmful chemicals from medical supplies in the US. But the road ahead is still bumpy. It can be difficult to actually figure out what chemicals are contained in the materials and components the company purchases from other manufacturers. Suppliers might have gone through numerous owners through the decades, and sometimes information about what materials they're using has been lost over the years. Not all of B. Braun's suppliers make parts exclusively for medical use, so they may not be accustomed to the same regulations or keeping track of what's in their products. When suppliers can't provide lists of the chemicals they're using, B. Braun does its own testing or splits the cost of testing through a third-party lab with the supplier to reverse engineer what chemicals went into a given product, which can be expensive and time consuming.

Once they've tracked down a comprehensive list of the chemicals in their source materials, the work of finding substitutions comes with its own host of challenges. Lots of research is required to avoid "regrettable substitutions"—swapping out one chemical for another that later proves to be just as bad or even worse than the chemical it replaced, as has happened with BPA and PFAS (per- and polyfluoroalkyl substances) in other industries.

New replacement chemicals must meet numerous regulatory requirements and pass extensive trials to ensure that they'll work with the existing manufacturing equipment and that they're compatible with processes further downstream like assembly, sterilization, and packaging. If a potential replacement chemical doesn't meet all the right requirements, they may have to go back to the drawing board to look for a new one that does, or investigate the cost and feasibility of changing their manufacturing processes to work with the new chemical. If the new replacement chemical does meet all those requirements, they move on to "functional specification and biocompatibility tests"—making sure the finished products work safely, as they're intended to, with the medications or solutions they'll be used for. And finally, they must seek regulatory approval for the new products using the replacement chemical—which is its own complex, multiphase process. So while there may be both ethical and business imperatives to get harmful chemicals out of medical supplies, it's not hard to see why most companies would just stick with what they've been doing without a regulatory requirement to implement these kinds of changes. And it's not hard to see why, even when regulations are enacted, it can take decades to actually see harmful chemicals entirely removed from the market.

Environmental and health advocacy groups are working with lawmakers to advance safer chemical laws for medical devices in the US. In 2021, for example, a group of six US congresswomen published a letter calling on the FDA to update its guidelines on the use of phthalates and endocrine-disrupting chemicals in IV bags and other medical equipment, recommend rules needed to protect patients from toxic exposures in medical products, and establish an education program to build clinician awareness about the risks of using medical devices with toxic additives like DEHP.[14]

In the meantime, numerous programs are working directly with hospitals, health systems, or doctors that want to reduce their patients' exposures to harmful chemicals, including Healthcare Without Harm, Practice Greenhealth, and the Alliance of Nurses for Healthy Environments.

Group purchasing organizations also have a role to play. Many GPOs have sustainability officers who spend much of their time sourcing safer products from companies like B. Braun and working to educate buyers about why it's worth switching over. "Alternatives don't always exist, so sometimes we work with suppliers to develop alternatives," a representative at one large GPO told me in 2021, pointing to the example of a PVC-free breast pump a supplier had developed for them upon request that's now widely available.

Sustainability directors at hospitals and hospital systems have some ability to push for safer chemicals too, but it can be an uphill battle. While it can be relatively straightforward to replace waiting room furniture and cubicle curtains that contain carcinogenic flame-retardant chemicals, advocating for phthalate-free IV bags or safer products for the neonatal intensive care unit requires buy-in from doctors, nurses, and purchasers—and sustainability directors are also charged with overseeing everything from climate-friendly hospital food to reductions in energy use, so they're often short on time and resources. "I spent some time connecting with physicians concerned about chemicals of concern in medical supplies, but we couldn't make headway at the time because the cost differential was so great for alternatives," the sustainability director of a large hospital system in California told me in 2021. "The challenge is, do those physicians spend their time lobbying at the federal level to change policy or work on changing internal practices? They only have so much bandwidth, so if they're focusing on policy changes, we don't have anybody to help manage internal and operational changes."

The good news is that while US regulations continue to lag behind, the market for medical supplies is rapidly shifting to favor healthier products. In 2020, B. Braun responded to material list requests for 1,200 products from its customers. In 2021, they responded to requests for more than 4,500 products. "The demands of our customer base are changing right along with the trends in green chemistry and sustainability," John said. "More and more customers are demanding that we be conscientious about these things, which we're in a good place to do—a better place than our competitors are."

B. Braun, like most large corporations, is neither all villain nor all hero. It's a company that mostly does good in the world while still prioritizing profitability, and that's striving to do better in some ways while falling short in others. These kinds of complexities and mixed incentives are not uncommon in global markets for any type of product. But when it comes to medical supplies, relying on the good graces of multinational corporations is not enough. The most vulnerable among us—premature babies fed through plastic tubes in the neonatal intensive care unit, chronically ill people who regularly receive IV medication through plastic IV bags, cancer patients like Berry, and children growing up near facilities that manufacture life-saving medical equipment—deserve the most protection we can collectively offer.

CHAPTER 6

Melanie: Safer Neighborhoods through Activism

When Melanie Meade was a kid, her mother used to call her and her siblings inside from playing in the yard every day at noon to make them change clothes and wash their faces because they'd be covered in soot. The soot came from what everyone called "the mill"—a vast network of buildings and smokestacks along the Monongahela River in Clairton, Pennsylvania, visible just down the hill from the family's house.

"The mill" isn't technically a mill. It's a plant that converts coal into coke, a fuel and additive used in steelmaking, by baking it at high temperatures. It isn't just any coke plant—it's the largest one in the United States, producing about 4.7 million tons of coke annually. It's also one of the oldest. The plant was built in 1901 by the St. Clair Steel Company before being acquired by the U.S. Steel Corporation in 1904 and renamed the Clairton Coke Works.

Melanie, now in her late forties, is slight with high cheekbones, dark brown skin, and a wide, bright smile. She often wears brightly colored silk head wraps paired with matching earrings, hiding the few strands of gray hair that belie her youthful appearance.

The first time we met was in 2019 at a press conference in Clairton, which is about 15 miles south of Pittsburgh. I was covering the event as a journalist, watching a handful of the approximately 36,400 people who live within 3 miles of the Clairton Coke Works demand action from local regulators following a string of "code red" air quality days when the air pollution was so thick it was unsafe to spend time outdoors. One of those days Melanie had gone to the emergency room with chest pain, afraid she was having a heart attack (it was a false alarm). She didn't want her son to play outside because she worried about what the air would do to his young lungs. When it was her turn to address the crowd, she looked directly into the local news cameras and spoke loudly and clearly. She was calm, but everyone in the room could feel her underlying anger. "I want to ask the health department, how long should our kids be playing inside? And more importantly, why should they have to?" she said. "We must not think that just because the air here is better than it was years ago, it is actually better."

Melanie was referring to the plant's early days, when the air in the surrounding areas—including Pittsburgh—was so famously dark with soot that streetlamps were lit in the daytime. A long-standing refrain among local politicians says residents should be grateful the air isn't that bad anymore. Thanks to the federal Clean Air Act, the air in Clairton is cleaner today, but U.S. Steel is still regularly fined and sued for illegally high emissions. The company has at times racked up several thousand air pollution violations in a single calendar year, has been fined more than $9 million for violations at the Clairton Coke Works since 2004, and is regularly engaged in numerous lawsuits with environmental advocacy groups and local residents. Activists have long complained that the fines U.S. Steel faces are so small compared to its net worth—$5.25 billion as of 2022—that it's cheaper for the company to "pay to pollute" than it would be to clean up its operations.

Melanie is a well-known spokesperson for clean air advocacy in

Clairton. She has helped organize countless community meetings and rallies, pleaded on the local news for health officials to do something about air quality, spoken about environmental justice at conferences and in meetings throughout Pennsylvania, and told her family's story in public testimony to local, county, and state politicians, including Pennsylvania governor Tom Wolf.

Melanie's family has deep roots in Clairton. Her paternal grandmother was born there, and her family, a long line of farmers and carpenters, owned farmland near the river long before the coke plant was built. When her paternal grandfather was young, he moved his family from their farm in Newport News, Virginia, to Clairton to find work at the mill. Since then, most of the men in Melanie's family have worked at the coke plant at some point in their lives. Some, including both of her grandfathers, had entire careers there, while others, including her father and brothers, worked summer jobs at the plant during high school and college. Some of the women did too—she remembers her aunt Betty working there for a time as a cleaner.

"It always seemed like they had the lowest-level jobs, and I don't think any of them would have felt welcome in the mostly white union," Melanie told me in her living room on a rainy, December day in 2021. "They didn't talk much about it, but they all seemed to feel they had some kind of duty to work there."

The Meade family home is a three-story brick house with mustard yellow double front doors. It sits at the end of a quiet, residential street on top of a hill overlooking the coke plant, on a property Melanie says was once unofficially reserved for white bosses at the mill. Beside the house, an old willow tree shades a picnic table. The sizable yard is lush green in the summertime, and a brick planter out front sprouts tall grasses and a few flowers. In the front window there's a small white sign with blue text that reads, "Clean Air Now!" The house belonged to Melanie's grandparents. After they passed away, Melanie's family moved in when she was in kindergarten. The door frame beside the refrigerator in the kitchen is crowded with the names of kids, scrawled in crayon beside notches marking their heights and ages.

Melanie's dad, who died in 2013, had a penchant for rehabbing discarded objects. The house is filled with stained glass windows, pews, floorboards, and door frames that he rescued from local churches slated for demolition. He also loved old things: African sculptures and masks from his trip to West Africa; antique dressers, secretary desks, and cedar chests in pristine condition; an antique barber chair leaking straw stuffing that Melanie remembers playing on as a kid. "Several people have expressed an interest in buying it, but the grandkids love playing on it and it brings back happy memories for me, so I haven't been able to part ways with it," she said.

When you drive into Clairton, you're greeted by a sign that reads, "Welcome to the city of Clairton, city of prayer." There's a church of a different denomination every few blocks—some blocks even have two—but many are now shuttered, along with most of the town's businesses. A sign reading "DANGER: DO NOT ENTER" hangs askew in the window of the long-abandoned hardware store, and the windows of a once-grand hotel that housed visiting steel barons have been boarded up, as have the doors of the former Union Trust Bank. If you live in Clairton, it's easy to get used to the stink that hangs over the town most days, but when you drive in from somewhere else, you notice it as soon as you step out of your car—a mixture of rotten egg smell from sulfur dioxide emissions, burning coal, and more acrid, hard-to-define industrial chemical smells emanating from the coke plant.

In the 1950s, Clairton's population peaked at 19,652, corresponding with the growth of the mill. In the 1970s the global steel industry sharply declined, and in 1983, U.S. Steel cut more than 15,000 jobs across its operations, marking the beginning of decades of decay for steel towns across the region and the nation. Clairton has lost more than

66% of its population since its peak, dropping to around 6,000 today. It has a poverty rate of 21%, significantly higher than the national poverty rate of about 11%. In the 1980s the Clairton School District consolidated all of its elementary, middle, and high schools into a single building. Plumes of smoke rise unceasingly from the plant's smokestacks directly behind the school playground. Today, about 1,200 Clairton Coke Works employees are represented by the local United Steelworkers union, meaning they live in surrounding areas, but fewer than 55 of them actually lived in Clairton proper as of 2019, according to the head of the local steelworkers union.

Melanie's parents, both born in Clairton, were high school sweethearts. Her mom, Maxine Pipkins, was petite and shy and grew up in a housing project near the plant; her dad, Thomas Meade, was a tall, outgoing football player from the wealthier part of town on top of the hill. Both of them came from big families. Her dad was one of nine kids and her mom was one of ten. They had Melanie's older sister, Tammy, when they were still teenagers in 1959. Her older brother, Michael, arrived in 1965. Melanie wasn't born until 1974, and her younger brother, Thomas Jr., was born four years later, in 1978.

High school football has always been a big deal in Clairton. The local team, the Bears, has won many state championships and once set a state record for the most consecutive wins (sixty-six). "Go Bears" signs are plastered all over town. Melanie's dad had earned a football scholarship at a local university and was on track to go professional, but he worried about the strain the sport put on his body and the headaches he got after taking hits, so instead he pursued a career in academia. He earned a doctorate in psychology at the University of Pittsburgh, where he became a professor in the school of education.

"He looked at education and getting a degree as a way to disprove a system that constantly told Black people that they were not very smart, that they were savages, that they couldn't achieve the things white people could," Melanie said. "It was a kind of silent revolution for him—it wasn't about throwing himself out there with pickets, it was about quietly getting to a place they said he could never reach, realizing that dream for himself and encouraging others to do the same."

While she was growing up, Melanie's family was considered well-off. They lived in that big house on the hill and regularly traveled to places like Virginia, North Carolina, and Myrtle Beach for family reunions and vacations. "I think those trips really broadened my perspective and shaped my thinking," Melanie said. "It made me realize things aren't like they are here everywhere. A lot of people here don't ever leave, so this area is all they know."

The Clairton of her childhood felt tight-knit. "You could see someone struggle and you'd see the community rally around them," she said. "Everyone had very little, but they still made sure everybody had something to eat. People kept space for others. They held a seat for you for the meeting."

Melanie's childhood was also sheltered. People looked out for her and her siblings because of her dad's position in the community. He was a lifelong member of the NAACP, he regularly volunteered at the local food bank, and he was heavily involved in local politics—he served two four-year terms on Clairton City Council, was appointed deputy mayor, and ran for mayor three separate times while Melanie was growing up, though he never won. He also grew a large garden in the empty lot across the street. It wasn't quite a community garden since Tom Sr. was very particular and didn't like anyone else entering his patch of Earth. But he enjoyed stopping kids as they walked by to ask if they'd ever tried a fresh tomato. At the time it had been decades since Clairton had its own grocery store, and when a passerby confessed to only having tasted tomatoes from a can, Tom Sr. would holler at Melanie to run fetch a knife, pluck a tomato from the vine, hand them a glistening slice, and watch their delight. He would often fill a paper bag with vegetables to send home with the neighborhood kids, and he tasked Melanie with walking around town delivering bags to elders who had trouble getting out to shop and offering them to anyone else who might be hungry. "I learned a lot from him," Melanie said. "He believed there was always opportunity for positive change. He never gave up hope for his community or his people." Melanie tries to apply that kind of thinking to her activism, but it's an uphill battle.

Emissions from coke ovens, like those in Clairton, contain particulate matter and a host of chemicals, including sulfur dioxide, formaldehyde, cadmium, and arsenic. Exposure to those emissions is a risk factor for chronic obstructive pulmonary disease (COPD), heart disease, and asthma, and the emissions are also potent carcinogens.

In 2005, the US Environmental Protection Agency (EPA) issued new regulations for coke ovens, but agency representatives admitted at the time that they didn't actually know whether the regulations were adequate to protect the health of communities nearby. The agency pledged to further review the matter, but it never followed up. In 2019, environmental groups, including Earthjustice, Citizens for Pennsylvania's Future, and the Sierra Club, filed a lawsuit against the EPA for ignoring requirements to update the emissions standards, specifically citing ongoing pollution from Clairton Coke Works. In 2020, the agency conceded in court that it had failed to properly regulate certain aspects of coke ovens, and it was given until 2023 to perform a risk assessment and update technology standards for controlling emissions. The local county health department, which oversees air quality for Pittsburgh and Clairton, recently attempted to enact more stringent regulations—in part due to pressure from Melanie and other community health advocates—but U.S. Steel successfully fought them in court so that only some of the proposed changes were made.

In the meantime, residents in Clairton still wipe thick coats of black soot from their windows, and students at Clairton Elementary school have asthma at more than double the national rate. The population of Clairton is small so analyses of its cancer rates are scarce, but research has shown that Allegheny County (which encompasses Clairton and Pittsburgh) is in the top 2% of counties nationwide for cancer risk from air pollution,[1] and the seven counties that make up southwestern Pennsylvania have substantially higher than average rates of several types of cancers with strong links to toxic chemicals, including leukemia, bladder cancer, breast cancer, kidney cancer, lung cancer, and thyroid cancer.[2] Residents of Allegheny County are exposed to coke oven emissions at a rate 88% higher than Pennsylvania's population as a whole and 99% higher than the US population. For the average American, the risk of getting cancer from breathing in coke oven emissions—assuming they have no other exposures to carcinogens—is 0.10 in a million. For residents of Allegheny County, it's 11 in a million.[3]

The Clairton Coke Works isn't the only industrial polluter in the region, or even the worst—it is the third-most-polluting facility in the county, bested by a company called ATI Flat Rolled Products Holding, LLC, which manufactures steel and stainless steel products, and Harsco Metals, which processes waste and materials from steel-making and other metal manufacturing companies. Numerous other metal processors and industrial manufacturing companies contribute to the region's air pollution, in addition to a vast oil and gas presence in rural parts of western Pennsylvania above the Marcellus Shale play. The new Shell petrochemical plant that converts ethane from natural gas into plastics is about 45 miles northwest of Clairton.

Melanie sometimes thinks of the Monongahela Valley, which is home to Clairton, as "cancer valley." She has watched more than her fair share of loved ones battle the disease.

The first was her uncle. It was 1992 and Melanie was eighteen years old and had just started college at the University of Pittsburgh. Her dad's oldest brother had worked at the coke plant for years before moving out of town. He returned to Clairton and told the family he thought something was wrong with him. Melanie remembers his head and neck shaking in a way she found alarming. Eventually he was diagnosed with metastatic cancer that had spread to his brain and progressed to an untreatable point. He had been a carpenter and a bricklayer in his later years, but his cancer made him unable to work. He didn't have a spouse or any children, so Melanie and her family walked down the hill to the house he was renting to help out until he needed full-time care, at which point they set up a hospital bed for him in their own dining room. "Others might have said he should go to a nursing home, but that was not going to happen because that's not how my family was," Melanie said. "He wanted to die with his family. So he was going to stay here and we were going to take care of him."

The experience changed Melanie's perspective. "When his time of death came, I was devastated," she said. "I worried there was more I could have done to save him, but I remember the hospice nurse pulling me in as he was taking his last breaths and saying, 'You weren't called here to stop his death, you were called here to ease his process of dying.' That helped me realize that I could handle this, and helped me recognize that having someone trust me to be with them at their time of dying was very profound."

Melanie would need those lessons later. In 2015 she cared for her older sister, Tammy, as she lay dying of lung cancer at the age of fifty-five in the same dining room where her uncle had died. Over the years she has also watched her aunt Betty, her grandmother, and two of her grandmother's sisters die of cancer. Many families in Clairton have similar stories, and Melanie worries about the collective impact.

"There's so much trauma that happens when a family encounters its first cancer diagnosis," she said, "feelings of hopelessness and helplessness, feelings of loss. And you often don't get to have a feeling of closure before the next person is diagnosed."

When we sat down to talk in December 2021, Melanie's cousin, who was in his early fifties, was dying of cancer. He had stayed with her for a while but had left to be with friends. He had decided to stop chemotherapy because it made him feel awful. He didn't want to be in the hospital or in bed, so instead he was trying to enjoy the life he had left, spending most days fishing in the Monongahela River, throwing the fish he caught back in. He called Melanie on one of his bad days and told her that he couldn't stop having flashbacks to watching so many of their other relatives die of cancer.

In Louisiana's cancer alley, in Clairton, and in similar communities throughout the country, Black and Brown residents are disproportionately sickened by pollution. In the early days of Clairton Coke Works, Black workers were typically given the most dangerous jobs, working directly with the hot coke ovens, and according to Melanie, their families were limited to homes at the bottom of the hill, closest to the mill's pollution, in a valley colloquially referred to as Black Bottom. These practices, historically common at steel mills and other industrial sites throughout the nation, along with Jim Crow laws and subsequent redlining practices, have resulted in what are sometimes referred to as "toxic zip codes."

For example, the Monongahela Valley (referred to as the Mon Valley), a former steel corridor of municipalities from Pittsburgh to the West Virginia border, regularly sees spikes in air pollution from the Clairton Coke Works and other polluting facilities, leading to some of the dirtiest air in the country. The Mon Valley is home to Clairton and many communities like it. The municipalities in the Mon Valley have poverty rates ranging from 14% to 40%—significantly higher than surrounding Allegheny County's poverty rate of 11% and the state poverty rate, which is also about 11%.

While class is often a factor in environmental health, research has shown that Black Americans are exposed to more air pollution than white Americans regardless of where they live or their income level, and that Black Americans are exposed to 1.5 times more carcinogenic pollution than the population as a whole.[4] A number of Mon Valley communities have substantially higher percentages of Black residents compared to the state or region overall: Allegheny County is 13% Black, and Pennsylvania is 12% Black, compared to Duquesne, which is 51% Black, Braddock, which is 69% Black, and Clairton, which is 42% Black.

This landscape mirrors national trends. It's easy to find stories about toxic zip codes, the people of color who are disproportionately affected by their pollution, and the activists trying to clean them up in most major American cities and industrial manufacturing hubs. There are toxic zip codes in Detroit, Denver, Salt Lake City, Houston, Dallas, Baton Rouge, Chicago, and Cleveland, and in rural parts of the country where fracking, mining, and industrial agriculture release toxic chemicals into the environment. In many of these communities, local leaders spend decades fighting for clean air and water. Their battles are just as critical to the new war on cancer as those being waged by academic researchers, DC lobbyists, and green chemistry advocates—and they're often significantly more urgent.

In addition to singing with her band, Berry also performs on her own as an indie folk singer-songwriter. In a few lines of her song "The Little Creek," she pays homage to Oil City:

> In the little creek I'll wash my wounds
> and heal them from inside
> There's a little creek flowing right through me
> It's on fire from all the oil
> The heart a sunken fire, abandoned coal mine set aflame
> The creek snuffed it out

"The Little Creek" is on the second of Berry's two solo albums, released in the summer of 2021. She named the album *Beam On*, which came from the machine she stared at while undergoing radiation therapy every day. She'd be lying on her side facing the machine, and when it was switched on, a little sign reading "Beam on" would light up to indicate that the treatment was under way. "I used that name for the album to honor what I've been through," she said. "And with hope for what's coming next."

Around the time her cancer treatments were winding down, Berry's mom became increasingly sick. She had smoked for most of her life and had developed COPD, which had her in and out of hospitals for much of 2021, with Berry heading up most of her care. It was difficult, but in a bright spot, Berry had her final cancer treatment earlier than expected, on October 5, 2021, one year to the day after receiving her initial diagnosis.

On Instagram, she posted a photo of herself in a surgical mask, her tired eyes squinted in a smile, her hand gripping the rope of the brass "ring me when you're cancer free" bell mounted at the entrance to the cancer center.

"So you think you had a rough year, eh?" she wrote in the caption. "Maybe you did, but let me tell you a little bit about my year. . . . Being an immunocompromised cancer patient enduring chemotherapy during a global pandemic was about as bonkers as it sounds. But thanks to many dear, selfless friends, loved ones and neighbors I made it through the most trying time I hope I ever have to endure. . . . Lots of clichés come to mind: new lease on life, newfound gratitude, etc. etc. but GAWD DAMN PEOPLE. Whew, what a long, strange trip it's been. Thanks for the love and support, food, friendship and attention. I love you. Check your tits. Don't ignore your health."

Warm and caring comments flooded in, with at least one person writing, "Boob check added to the agenda. Thanks for the reminder in the form of this lovely post."

Finishing that final treatment was a huge relief, but Berry still couldn't shake the feeling that it wasn't really over. For one, she still had a port in her chest. The surgery to remove it was scheduled several weeks later. But more than that, she would never be entirely free from cancer. She would think of it every time she looked at the scars on her chest. She would fear its return whenever she had an extra drink with friends, got exposed to secondhand smoke, or needed a follow-up scan.

A month before her final treatment, Berry had gone in for her first postsurgery mammogram. Her right breast, the one she'd had cancer in, looked fine, but her doctor seemed worried after looking at scans of her left breast and requested additional imaging. The déjà vu was nightmarish, a cold trickle of fear running through her body.

The additional scans were done over a period of a few days, which Berry wandered through in a haze. She alternated between reassuring herself that everything was probably fine and spinning out on fears that she would have to start the whole terrible process over again. Maybe this was just how her life would be from now on—scan after scan, treatment after treatment, never reaching the other side of cancer.

In the end, it turned out to be nothing. She learned that she just has very dense breast tissue, which likely means she will have to endure a similar process for future scans. "Cancer survivors call this 'scanxiety,'" she told me. "Every time you go in for a follow-up scan it brings up all those feelings from the time you got your diagnosis. It's scary. I don't think it will ever stop being scary."

———

When she was in her twenties, Melanie had been living and working in Washington, D.C. for two years when she attended the Clairton Reunion, an annual tradition where locals who've moved away come home and get together over Labor Day weekend. The reunion was held at a bar near the coke plant, with grilling and dancing outdoors in a side lot. "The last memory I had was standing out there, enjoying the party, and then I woke up the next day in the hospital," Melanie said. "It turns out I'd had my first seizure."

Melanie was diagnosed with nocturnal epilepsy—a form of the disease where seizures typically happen at night, before bedtime, or shortly after waking up. Melanie's older sister Tammy had epilepsy, so she was familiar with the disease. Several studies have shown increased hospitalization rates for epilepsy following spikes in air pollution,[5] and Melanie is now convinced that being reexposed to pollution from the mill initiated her first seizure. "It started when I was here, right next to the mill," she said. "That diagnosis completely turned my life upside down."

The condition and the medication she was prescribed for it made it difficult to function. She was exhausted, lost her appetite, and had trouble caring for her older son, who was small at the time. It took her a while to find the right balance of meds and natural remedies to keep the seizures at bay but not completely knock her out.

In 2013, Melanie was managing her disease well and had moved to a small town in North Carolina, where she was working toward buying an apothecary and establishing her own natural medicine practice, when her father died of heart failure associated with COPD. When she returned to Clairton for his funeral, she planned to help handle his estate and spend some time grieving with the rest of her family before returning to her life in North Carolina. But while she was in town, her mom's COPD took a sharp turn for the worse, and Melanie took over caring for her. Six months later her mom died too.

It was not long after, while Melanie was wading through the fog of grief, when she first learned about how bad the air pollution from the coke plant was. A local chapter of the Clean Air Council held a community meeting and shared that Clairton had high rates of both cancer and asthma, which they said were linked to the plant's emissions. Melanie's younger brother Tommy had childhood asthma. It had been terrifying for Melanie to witness him having asthma attacks and being rushed to the hospital, but she had never before made the connection between her family's ailments and the coke plant's ever-fuming smokestacks.

Both of her parents smoked, which Melanie knows was the primary culprit for their health issues. But she was surprised when others in the community dismissed the idea that pollution from the mill could also have played a role. "It seemed like there was an unspoken rule that everyone had to have loyalty to the company," Melanie said. "I have never understood it."

It's a pattern she sees a lot. When friends and relatives develop cancer, they tend to blame themselves. "They'll say 'oh, it's because I didn't exercise enough, or I didn't eat well,'" she said. "It's been ingrained in them that it's not anyone else's responsibility."

She pointed to a 2021 analysis in which researchers calculated that even if everyone in Allegheny County had quit smoking twenty years ago, the region's lung cancer rates would only be 11% lower.[6] Among the 612 other US counties in the study, lung cancer rates would have declined an average of 62% if smoking hadn't been part of the picture over the last twenty years, indicating that environmental exposures play a substantial role in the region's lung cancer rates. "I'm not trying to tell people to ignore their own responsibility," Melanie said. "But it's very damning to a person's spirit to think they're at fault for their sickness, and that they failed by not being able to control their own health."

After both her parents died, Melanie started attending more clean air meetings, finding that she felt better when she occupied her mind. At one such meeting she learned about the study that found kids in Clairton have asthma at more than double the national rate. "That's when I decided I really had something important to do here," she said. "Who could hear that and not want to scream and try to get Oprah on the phone and demand change?"

Before long she was speaking at meetings and sharing her own experiences with living in Clairton: the foul-smelling air, her struggles with epilepsy, the illness in her family, and her belief that everyone deserves a healthy environment. She still had the little apothecary in North Carolina in the back of her mind, but she wasn't quite ready to go. Advocating for clean air in Clairton felt like her purpose—a continuation of her dad's dedication to the community. And staying in her parents' house made them seem still close by. As it turned out, any idea Melanie had of returning to her old life was short-lived.

The same year her mom died, Melanie's oldest brother, Michael, died at age forty-five from congestive heart failure associated with COPD. Two years later, in 2015, her older sister Tammy died at age fifty-five from lung cancer. And in 2020, at age forty-one, her younger brother Tommy died from a blood clot in his lungs that may have been related to COVID-19. In the span of eight years, Melanie had buried every member of her immediate family.

Living in a more or less perpetual state of grief taught Melanie a lot about trauma, self-care, and resilience. There were times it felt like she was drowning. She sought therapy and briefly took antidepressants, but in the end she wound up relying more heavily on spiritual guides and community connections to heal. Now, though there's still sadness about losing so many people she loved, Melanie thinks of death more as a shifting spiritual state than an absolute end. Death has become so familiar that she's considering certification as a death doula—someone who coaches the dying through their journey out of this world, much in the same way a birth doula coaches parents on bringing new life into the world. Unlike hospice workers, death doulas deal less with the physical aspects of dying and more with the emotional, mental, and spiritual ones. Above all, they advocate for their clients' own wishes about dying, even if they're different from what traditions (or pushy family members) dictate.

When we spoke in 2021, Melanie had already started informally providing care for others who were dying in the community. She was with four Clairton community members outside her family at the time of their deaths too. "I think the traditional Western death experience leaves us feeling like victims, like we've suffered a loss and can't get it back, and that's not where I want to finish with death," she said. "I'm interested in revisiting some of our older death rituals that can help people tap into feeling more empowered when they die."

Coaching others through loss also taught Melanie how to keep going in her activism, even when it feels hopeless. "When I first started working in environmental justice, I felt awful when things didn't happen as quickly as I wanted them to, or at the magnitude I wanted them to. But now I'm learning that when things aren't going my way, I need to be able to find a place of peace. Sometimes when I stop looking for *my* way, I find that another way opens up."

———

Two weeks after her final treatment, Berry's mom died from complications associated with COPD.

It happened on the day that Berry finally had her chest port removed. Her mom had already been in the ICU for several days, and she wasn't doing well. Her brother wanted Berry to reschedule the surgery so she wouldn't have to leave their mom, but she just couldn't do it. She'd been waiting too long for this, a moment of resolution amid all the stress and sadness of her illness, the pandemic, and her mom's declining health. "I needed that thing out of me," she said.

She kept the appointment, traveling to Pittsburgh for the surgery and driving back to Oil City afterward. In the place where her port had been, there was now a small bruise along with several stitches that would dissolve in the coming weeks. She felt a little lighter on the drive back. She kept fingering the place where her port had been through her sweater, reassured by the flatness. She arrived back in Oil City in time to see her mother before she died.

The next day, Berry posted a series of photos on Instagram: One of her mom in the 1970s in a mustard yellow turtleneck and blue denim overalls, young and round-cheeked and grinning; one of her mom in 2019 surrounded by her three children, Berry with long hair that fell all the way to her breasts precancer; and one of her parents in the 1980s when her mom had a short, feathered haircut that resembled present-day Berry with her postchemo short hair.

"This woman was a force of nature," she wrote in the caption. "She was dedicated, determined, and intelligent. She taught me the cardinal sin of painting original woodwork. She had impeccable style and expensive taste. My understanding of aesthetic principles seeped from her into me. We didn't always see eye to eye and we certainly didn't agree on politics and lived very different lifestyles, but we loved each other. I miss her already. It was horrible seeing her suffer, but she fought hard until we lost her last night. I love you, Mamma." She ended the post with a no-smoking-sign emoji.

Grief washed over Berry, but there was relief too. It had been hard to watch her mom suffer, and hard to spend time in the Oil City house with her brother and sister. To cope, they drank and smoked more than usual, both of which made Berry feel simultaneously tempted and triggered—they were some of her old favorite coping mechanisms too, but now she was terrified of their potential to make her cancer come back. Instead, she focused on the arrangements that had to be made—paying hospital bills, planning a funeral. She and her siblings pored through family photos, taping dozens of pictures of their mother onto pieces of poster board to display at the ceremony.

For weeks after her mom's death, Berry moved through her life in a fog, doing what had to be done but little else. It occurred to her that her mom had lived, maybe even waited, just long enough to see Berry finish her cancer treatments.

————

Melanie keeps an altar in her bedroom on the third floor of the family home, where she now lives alone. The room is technically the attic of the house, with slanted ceilings and numerous closets lining the walls. The altar rests on top of an electric fireplace with ceramic logs. It's covered in candles and photos of the dead—her mom and dad, all three of her siblings, friends, grandparents, aunts and uncles. Crystals are scattered among the photos: black obsidian, citrine, rose quartz, and amethyst, along with a blackened piece of palo santo and a few half-burned bundles of sage. There are also offerings: coins and dollar bills, a few oranges, a wine glass full of water, a tumbler full of whiskey. "In our culture we say 'pour one out for the homies,'" she said, laughing, when I asked about the significance of the whiskey. "It's just something they would enjoy."

On one corner of the altar there's a small, ceramic sculpture of a Black woman in a blue nightgown and hair rollers. She sits hunched over on the edge of a bed, her head hanging forward, with one foot in a slipper and the other out. Her smooth face has no features, but everything about her posture evokes exhaustion. Melanie showed me the bottom of the figure with its title: "Blue Monday."

"It was my mother's," she said. "It's always reminded me of how she seemed to feel—constantly tired. And it's always made me think that I didn't want to live like that, just living to work. I wanted something more meaningful. When she died, I didn't feel like I could get rid of this because it reminded me of her, but I also didn't want to look at it all the time and feel sad. So I put it here as a way of honoring the way life was for my mother, but also leaving those feelings with the dead, along with my grief—kind of giving it over to them to manage so I don't have to live with the burden of it."

Through the years of moving through grief while fighting for clean air in Clairton, Melanie was often frustrated by her friends' and neighbors' lack of support for the cause. She regularly came up against that same blind loyalty to the steel company she'd seen in her own family. That was especially true among the members of Clairton's city council, including the mayor, who had worked for U.S. Steel for more than thirty years, continued to work for the company while serving as mayor, and remained vocal about his support for the jobs provided by the plant, despite having had cancer twice himself.

Some think this phenomenon is driven by a willingness to sacrifice—the idea that communities like Clairton are willing to give up their own health, not only for access to jobs but for the good of the nation. It's an attitude that persists in places where extractive industries have always been dominant, like the greater Appalachian region where first there was coal, then oil, then coke and steel, and now natural gas. All have left death, disease, and environmental disasters in their wake. But dying of black lung or in a mining accident is often portrayed as an act of patriotism necessary to keep America running, similar to the sacrifices made by fallen soldiers or police officers. Casualties of our unsuccessful war on cancer are often bestowed similar narratives: heroes who battled valiantly, rather than victims of corporate greed and regulatory failure whose illness could have been prevented. U.S. Steel is a for-profit company whose CEO took home more than $10 million in total compensation in 2020. But for many Clairton residents, the company might as well represent American values themselves. And those values demand loyalty.

Finding those sentiments in many of her neighbors was maddening, but Melanie also found support in places she never expected to. "I heard from workers at the plant, total strangers," she said. "They weren't going to let anyone know publicly that they supported me, but they took time to reach out and tell me how much they appreciated me speaking out. And I heard from some wonderful women in other areas nearby that are also experiencing environmental injustice, who reached out to invite me to lunch and talk about what more could be done and how we could support each other."

A few years into her advocacy, Melanie began to see some signs of change. Clairton's pollution problems were getting more and more media attention, local health officials publicly promised to crack down on polluters, and state officials began to talk about the importance of environmental justice in campaign speeches. Progress felt slow, but imminent.

Then, on Christmas Eve in 2018, a major fire at the Clairton Coke Works took out the plant's pollution controls, resulting in severe, repeated spikes of emissions. The local health department found out about the fire that day. But most Clairton residents—including the mayor—didn't learn about the fire until January 9th, when the county issued a formal air quality alert following six days where pollution levels violated federal air quality laws.

"I was enraged," Melanie said. "I felt defeated at that point. I felt like everything I'd done was a waste and nothing would ever change. I wondered if I'd missed out on what my true purpose should have been, and if I should have given more time to the family I have left for all those years instead of bothering."

She thought about leaving. One of the first air quality activists she'd ever connected with told her it seemed obvious that nothing would ever change and that she should probably just move somewhere safer. "I promised myself I wasn't going to die here like my parents," she said. "But I think I made that promise as a reaction to my rage. It suddenly seemed like my parents gave up too much for the community they loved, but I was also meeting with myself in the mirror, because I knew I would have been giving the community false hope if they couldn't see anyone fight the fight until the end. I reminded myself that things often have to get worse before they can get any better."

U.S. Steel didn't finish repairs on the plant's controls until April 2019, and the mill continued pumping out vast quantities of pollutants in the meantime. A 2021 study found that asthma-related outpatient and emergency department visits nearly doubled among people living in the Clairton area in the two months following the fire.[7] A subsequent lawsuit argued that the blaze was caused by a series of systematic failures resulting from decades of neglected maintenance and lack of investment at the plant.[8]

Despite these frustrating setbacks, years of activism by Melanie and others like her have borne some fruit. In 2020, U.S. Steel settled a class action lawsuit claiming Clairton Coke Works' pollution created a nuisance and hurt nearby property values. The lawsuit didn't claim any health impacts, in part because there are so many other industrial facilities in the region that proving just one of them caused residents' health issues in a legal battle is nearly impossible. The settlement required U.S. Steel to spend $6.5 million toward environmental improvement projects at the facility and paid about $2 million to claimants, about half of which went toward attorney fees and legal costs. In the end, Clairton residents who didn't opt out of the lawsuit received checks for less than $300 and waived their right to sue U.S. Steel for nuisance or loss of property value that occurred prior to December 24, 2018.

Melanie opted out of the settlement. She didn't think $300 was worth it, and she said she wanted to hold out for a lawsuit that addresses health effects from the plant's pollution. She also didn't think the settlement was fair. "If there's $6.5 million in work that needs to be done so they can follow clean air laws, it shouldn't have to come out of this lawsuit," Melanie said at the time. "They should be doing that anyway. That money should be given to the people who lost out, not earmarked for the company to help itself."

The county health department, which oversees air quality, finalized a separate settlement agreement with U.S. Steel in 2020 that created a community trust for residents of Clairton and four other nearby communities affected by the plant's pollution. About $2.5 million in fines were put into the trust, but community advocates were frustrated that Clairton residents weren't prioritized despite being the most affected. They also complained about a lack of transparency and public involvement in the payout process, which empowered just one representative from each community to determine how the funds would be spent, with no requirement that they be used to alleviate the harms of air pollution. As of 2021, Clairton had put funds it received from the trust toward a new public works truck and establishing a community center.

In 2021, U.S. Steel announced it was reneging on a long-standing promise to invest $1.5 billion in equipment upgrades that would have lowered harmful emissions at its Pittsburgh-area plants while providing the region with up to 800 additional union jobs, opting instead to buy a more state-of-the-art, nonunion steel mill in Arkansas. As a concession, the company said it planned to shut down three of its oldest, most heavily polluting coke ovens at the Clairton plant in 2023.

The county health department also enacted a new policy aimed at preventing extreme air pollution days, particularly in communities like Clairton, in 2021. It requires seventeen local industrial polluters that affect air quality in the Mon Valley, including the Clairton Coke Works, to lower their emissions when forecasted weather conditions predict an "inversion," which occurs when a warm air mass sits above a colder air mass, trapping pollutants close to the ground. An inversion resulted in the worst air pollution disaster in US history—the 1948 "Donora Smog," which trapped industrial pollution in the Mon Valley town of Donora, just 14 miles south of the Clairton Coke Works Plant, killing twenty people and ultimately spurring the creation of the federal

Clean Air Act. Valleys are especially prone to inversions because the surrounding hills and mountains hem in air. Historically, inversions are rare, but climate change is driving an increase in the phenomenon across the globe, and they're expected to become more common moving forward—especially in places like the Mon Valley—so this new local law is important.

None of these things have been enough to fix the problem and none have happened in quite the way she had hoped they would, Melanie said, but any progress is valuable.

Melanie isn't working alone. Many local organizations, including Allegheny County Clean Air Now, the Breathe Project, the Group Against Smog and Pollution (also known as GASP), and the Cancer and Environment Network of Southwestern Pennsylvania (which was inspired by the national Cancer Free Economy Network), are advancing the fight for cleaner air in Clairton and the Pittsburgh region. Connecting with activists in other regions has also helped Melanie stay hopeful and devise new strategies. She has worked with the Black Appalachian Coalition (BLAC), an organization of community advocates working on various community issues; with a clean air group in Dallas called Downwinders at Risk; and with a group of researchers studying ways to improve the health of people living amid air pollution in Louisville, Kentucky. She has also connected with activists in Beaver County to offer support for their fight against Shell's petrochemical development and request support for her efforts in Clairton, and with activists in cancer alley, Louisiana, who helped her find perspective when she was feeling frustrated.

"They helped me understand why a person might not have the capacity to be involved, so I no longer had any negative feelings toward people for not turning up at meetings or rallies," Melanie said. "They helped me learn how to look within myself and to keep myself balanced, and reminded me this work is not for my own personal outcome, but for the benefit of the entire community, so working in a way that will let me stay the course and keep caring about the community is what's most important."

Melanie takes solace from organizers who win against big polluters. In Louisiana's cancer alley, for example, activists have stopped several new petrochemical plants from being built in recent years. Downwinders at Risk in Texas have pressured officials to properly regulate emissions from local cement plants, forced the closure of a lead smelter that was illegally disposing of waste in Frisco, and secured the passage of a stricter gas drilling ordinance in Dallas.

Melanie has also met with activists from a whiter, more affluent community in nearby Pittsburgh, who helped force the closure of a plant very similar to the Clairton Coke Works. The Shenango Coke Works, a highly polluting facility owned by DTE Energy, was shuttered in 2016 after years of community protest and citizen science showing flagrant clean air law violations. Local activists like Melanie worked with researchers at Carnegie Mellon University to set up cameras trained on the facility's smokestacks, which regularly provided images of emissions online. Residents learned to review the images for potential air pollution violations by checking for things like the opacity and direction of the smoke, flagging emissions that looked off so the time stamps could be compared with air pollution monitoring data. They watched the facility like hawks, flagging, recording, and reporting many violations. Following numerous fines and lots of bad publicity, DTE Energy, the company that owned the plant, announced it would close the plant, citing poor market conditions. A study by the local health department found that in the year following the plant's closure, emergency room visits for asthma and COPD dropped by 38%, and visits for cardiovascular diseases, including heart attacks and strokes, decreased by 27%.[9] It's still too soon to know how the closure will affect cancer rates. Carnegie Mellon University has set up a similar network of cameras trained on the Clairton Coke Works, and activists like Melanie have been dogged about reporting potential violations. Their efforts have also resulted in some additional fines and media coverage, but there's no indication that the plant will close.

Learning about successful activism in other places makes the fight in Clairton less lonely, Melanie said, like she's part of a global movement toward environmental justice instead of just one person in one community that's fighting for its survival. She also wants to start finding ways to create solutions for her community, in addition to always fighting the problems. She hopes to one day turn her house into a community center where people can learn to advocate for clean air and protect themselves from pollution in the meantime by running a box fan with a HEPA filter cut to fit in it, as Melanie does, and using air monitors to know when it's safe to spend time outdoors. She also hopes to grow food for the community like her dad did, and offer free meals made with fresh vegetables and medicinal plants.

"I want this house to be a place where the community can process all this trauma in a home where we have comfort and nurturing," she said. "Where people can have a warm meal and a soft place to sit and talk. . . . Those are the things my community doesn't have enough of."

Melanie still worries that living amid the pollution in Clairton could take years off her life. "I'm forty-seven right now, and I have three grand-children," she said. "I wouldn't have wanted them at this age, but now I'm grateful I have them early because most Black women here don't live past sixty. Just last year I knew of three women under the age of forty who passed away from heart disease, kidney failure, and cancer."

"That's also why I want to create something that's not going to die with me."

———

On a Friday night in December 2021, about eight months after I first met Berry, I stood beneath a disco ball in a basement bar watching her perform with her indie folk band.

She looked markedly different from the day we first met. Her hair, which had been straight her whole life, had grown in curly—a common phenomenon sometimes referred to as chemo curls—and had been shaped into a shaggy pixie cut. She looked luminous in a sleek, sleeveless black jumpsuit, turquoise jewelry, and bright red lipstick.

As the only woman in the band, Berry's voice cut through, soft but clear, creating ethereal two- and three-part harmonies with lyrics like, "What a holy grace I saw your body had as I watched you make for the water-hole, and your footing falls onto the heather soft as a sailor's prayer when he meets his death," and "And then I hold you. And you hold me. I can hear the future calling from a star above the sea."

Standing there watching her, it was impossible not to feel joy and gratitude on her behalf. In the last few years, she had been diagnosed with cancer, then endured multiple surgeries and twelve months of chemo and radiation that ravaged her immune system during a global pandemic. She had been at turns too sick and too terrified to leave her house. Her mom had been in and out of the hospital half a dozen times, and then had died, just as Berry was finishing her cancer treatments.

But here she was, standing on stage, singing sweet harmonies with old friends.

The crowd was filled with flannel, beer, and nodding heads. I might have been the only one who knew what Berry had overcome to arrive here. But I imagined her voice igniting little sparks of hope and gratitude in everyone who heard it, even if they didn't know why.

Moving Beyond Survival

While millions of people participate in various "races for a cure" that raise millions of dollars for cancer research every year, only a fraction of those funds go toward prevention. Races for prevention are rare because there's no survivor to embody success or thank us for our efforts.

"Curing someone or being cured is viewed as a personal triumph, so everyone is focused on treatment," said Dr. Margaret Kripke. "Prevention doesn't have a face. We don't get to know whom we prevented cancer in, which is why it's easier to advocate for the fight for a cure than it is for prevention."

Kripke is the cancer expert who served on the President's Cancer Panel, which found that up to two-thirds of all cancer cases are linked to preventable exposures. In 2014, while Kripke was serving as chief scientific officer for the Cancer Prevention and Research Institute of Texas, she herself was diagnosed with breast cancer.

"No one in my family had ever had breast cancer as far as I knew," she told me in the summer of 2021. "Like so many people developing cancer these days, I wondered, how did I get it? Where did it come from?

Was it just bad luck? A mistake in DNA replication, or some causal event during childhood or even in utero that triggered that?"

Kripke felt lucky to understand as much as she did about cancer when she became a patient—she knew her prognosis was good, and she felt confident in her doctors' ability to successfully treat her. After surgery and six months of chemotherapy, she was in full remission. "I wasn't terribly anxious about it, but it was very unpleasant," she said. "I certainly would have rather had prevention than cure."

Efforts are under way to get big cancer nonprofits like the American Cancer Society, the American Childhood Cancer Organization, and the Susan G. Komen foundation, among many others, involved in the fight against carcinogens in our environment. The Cancer Free Economy Network is working to help organizations that serve cancer patients update their educational materials with information about the role of toxic exposures. And advocates with the Healthy Building Network are pushing organizations that provide long-term housing to cancer patients and their loved ones to build them without the use of materials containing carcinogens. If initiatives like these are successful, maybe someday a race for prevention won't seem so far-fetched.

In the first phase of our "war on cancer," we put nearly all our resources into fighting for a cure—treating and healing soldiers who've been wounded on the battlefield, if we stay with the war metaphor. Some good things have come from focusing on treatments and cures. Scientific advances have led to a steady increase in the number of people who survive cancer. The overall death rate for all cancer types in adults declined 32% between 1991 and 2019, and many more children survive cancer today than in the past.[1] Survival rates for both childhood leukemia and childhood brain cancer, the two most common forms of cancer in children, are around 85% and 70%, respectively.

eenwashing, a precursor to decades of marketing spin by
ustries.[2]

companies have followed a similar blueprint, pushing per-
sibility for the climate crisis through the idea of a "car-
t." British Petroleum, or BP, invented the term in the early
iling a "carbon footprint calculator" that encouraged con-
ink about how their daily behaviors were responsible for
lobe. Meanwhile, since the 1980s, just 100 fossil fuel pro-
uding BP—have been responsible for about 70% of global
ning emissions. BP is near the top of the list of the high-
companies in the world, responsible for more than 34 bil-
ons of carbon emissions since 1965. In marketing materials,
urged consumers to consider going on a "low-carbon diet"
its own forays into renewable energy, while investing less
f its 2018 budget in renewables. Other companies jumped
wagon and started pushing consumers to mind their car-
t—a concept that took deep root in our public discourse.
rm is everywhere, used by the likes of the US Environmen-
n Agency, the *New York Times*, and countless brands.[3]

tics seem especially egregious given that fossil fuels and their
re responsible for much of our daily exposure to carcin-
include air pollution from the burning of fossil fuels to
rgy, vehicle emissions, petroleum-derived pesticides, dozens
n-derived carcinogenic ingredients that make their way into
products, and chemicals that can leach into food through
aging. The companies profiting from the continued prolifer-
il fuels and plastics are not only driving climate change but
uting to cancer cases across the globe.

magine these corporate titans and PR execs, it's hard not to
rtoon villain cackling maliciously over his piles of money.
the reality is much more mundane: short-sighted decisions

Madelina, the little girl from the introduction whose photo went viral, is among the survivors. When I first spoke with her mom, Kristin, in 2019, Madelina was a healthy, happy-go-lucky kindergarten student who loved ballet and gymnastics, riding her bike, swimming, and playing outside with her older sister. No one would ever look at her and guess that she'd been diagnosed with cancer before even reaching her second birthday.

But even when treatment is successful, it's never easy. Children often lose their hair and endure side effects like nausea, vomiting, pain, behavioral problems, and anxiety. Just as Berry feels that she'll always be haunted by breast cancer, the effects of enduring childhood cancer often last a lifetime—for both survivors and their families.

"Surviving cancer is life-changing for the patient and the family," said Dr. Erika Friehling, a pediatric hematologist/oncologist at the University of Pittsburgh Medical Center's Children's Hospital, where Madelina received her treatment. "Even seeing patients that completed treatment five to ten years ago, they still carry many of those emotions. Some parents talk about a kind of PTSD phenomenon every time they come back to Children's Hospital."

Kristin said she gets that—she and her husband find themselves holding their breath as they await the results of Madelina's regular cancer screenings, overwhelmed with gratitude each time one comes back negative.

In addition to the emotional strain, there's a significant physical one: 60% of children who survive cancer experience health problems related to the disease or their treatment, including infertility, heart failure, and secondary cancers later in life.

"We're also learning more about the neurocognitive impacts of undergoing chemotherapy as a child," Friehling said. "Specifically for the treatment of acute lymphoblastic leukemia, we're seeing some struggles in learning and developing more complex processing skills as kids get older, so we're beginning to put an emphasis on making sure those patients get extra attention as they're being treated or after being cured."

Kristin makes sure Madelina gets tested for heart problems and cognitive issues regularly since completing her treatment. "There are a gazillion unfortunate side effects," she said. "Fortunately we haven't encountered any other than some very minor developmental delays, but as she continues to grow, they're definitely something we need to watch out for."

As a result of both the growth in childhood cancer and ever-improving medical advances, one in every 530 American adults ages 20–39 has survived childhood cancer—and each faces the potential for lingering side effects for the rest of their lives. And that's just childhood cancer survivors. The American Cancer Society estimated that there were 16.9 million cancer survivors in the US as of 2019, or about 5% of the total population, and that by 2030 the total number of American cancer survivors will increase by more than 31%, to 22.1 million.

Cancer survivors are often encouraged to eat "clean" and avoid chemicals in their daily lives to help prevent their cancers from coming back. This can help some survivors feel empowered to protect themselves and may improve their health and future prognoses. But it can also make some survivors feel an unwarranted sense of guilt over the role their consumer choices may have played in their development of the disease, or feel a sense of failure if their cancer does return. And no matter how carefully survivors shop, it's just not possible to avoid carcinogenic chemicals entirely, because they're ubiquitous.

True empowerment would mean everyone being equally protected from exposure to cancer-causing chemicals. A successful war on cancer would do more than just increase the odds of survival—it would reduce the number of people ever forced to battle the disease. More than fifty

years into this war, it is long past requires going on the offensive.

The "Personal Responsibility" M

The idea that our individual choice health of the planet—and in turn take hold by chance. Companies plastic packaging made concerted dealing with cancer-causing waste i effectively letting industry off the h

In the 1950s a group of comp more heavily on plastic packagin and Anheuser-Busch (among other America Beautiful." Its purported mental stewardship, but its ulterio for plastic waste from the compa In 1971 with help from the Ad C out its infamous "Crying Indian" "Indian"—actually an Italian Amer hair—crying over the sight of trash service announcement was, in fact leveraged the American public's co tory of violence against Native Ar new, disposable packaging generat large corporations littered the envi blame. Meanwhile, behind the scen would have made plastic producers measures that could have ameliora now been escalating unabated for r Beautiful's initiatives have been de

corporat
polluting

Fossil
sonal res
bon foot
2000s, u
sumers t
heating t
ducers—
climate-
est-emitt
lion metr
the comp
and tout
than 2.3
on the b
bon foot
Today, th
tal Protec

These
derivative
ogens. Th
generate
of petrole
personal
plastic pa
ation of f
also contr

When
picture a
But I kno

made by many corporate employees who are doing their jobs exceptionally well, maximizing shareholder profits with little regard for anything else. As much as corporate greed, the consequences represent profound regulatory failure—ultimately traceable to a deeply flawed political system without effective campaign finance laws. If citizens are responsible for the carcinogens in our environment, it's not because we made the wrong consumer decisions, but because we—collectively—failed to properly govern.

Shifting Our Attention to Create Change

The idea that our everyday habits alone determine our cancer risk is similar to the myths about individuals bearing total responsibility for the plastic and climate crises. Our choices as individual consumers do matter when it comes to our health, but we can't solve these problems by changing those behaviors alone. We have to change our regulations to protect everyone from harmful chemicals, minimize the influence of large corporations over policymakers, and force corporations to pay for the true cost of their operations instead of continuing to externalize them at the expense of our planet and our health.

Hundreds of thousands of people around the world die from lung cancer caused by air pollution annually, and it's estimated that more than 100,000 lifetime cancer cases in the US alone could be attributed to carcinogens in tap water. And pollution in air and water account for only a fraction of the harmful exposures we experience every day. How different would our world look if we could put names and faces to the hundreds of thousands of lives that would be saved if just those two sources of exposure were eliminated through protective policies and regulations?

There are lots of ways to effectively push for meaningful, systemic change. Here are a few:

- Donate time or money to organizations that are already engaged in this fight (see the appendix for a list).
- Follow these organizations on social media, and answer and amplify their calls to action in the form of petitions and outreach to lawmakers about pending legislation.
- Sign up for notifications about proposed federal and state regulations related to carcinogens in the environment so you can contact decision makers about them when they're up for consideration (some websites where you can do this: govtrack.us, congress.gov, openstates.org, ncsl.org).
- Regularly remind your elected officials that this issue is important to their constituents.
- Amplify important stories about getting carcinogens out of the environment by sharing them and talking about them.
- Create market pressure by communicating to corporations that this issue is important to consumers. For example, you might tell your favorite microwave popcorn manufacturer that you have stopped buying their product because of concerns about PFAS in their popcorn bags, or tell a makeup brand that you've switched to their products because you appreciate their commitment to using safe ingredients.
- Support efforts aimed at campaign finance reform, restricting the power of lobbyists, and increasing transparency in the US government. (Organizations leading this work are also included in the appendix.)

If we all take even a sliver of the energy we devote to lowering our individual carbon footprint, being diligent recyclers, and reducing our personal cancer risk, and channel it toward implementing systemic changes that will protect everyone from cancer, we'll get a lot further a lot faster.

"In the end, the thing that usually gets harmful chemicals out of the public domain is public outcry," Kripke said. "Public awareness is really increasing about the fact that there are bad things in the environment we could get rid of that might make a difference right now. Knowing that gives me hope for the future."

Acknowledgments

I'd like to thank the scientists, researchers, activists, doctors, parents, and advocates who strive every day to make the world a safer, healthier place, and who generously shared their insights and ideas for this book.

Thanks also to my early readers for this book, Hattie Fletcher, Polly Hoppin, Nick Krieger, Jon Meck, and Pete Sheehy, whose feedback was invaluable. Thank you for the time, care, thought, and attention you devoted to my work, and the thoughtfulness with which you shared suggestions. My early interviews with Polly Hoppin about the Cancer Free Economy Network for *Environmental Health News* stories sparked many of the epiphanies that made me want to write about cancer and the environment, and despite being a powerhouse with a perpetually full plate, Polly has always been generous with her time, for which I am profoundly grateful.

I owe many thanks to my husband, Michael, whose love, support, and expert cooking skills have nourished me enough to make this project possible, and to our dog, Mochi, whose cuddles and shenanigans make every day better. I also owe many thanks to my family, who have always encouraged me in my writing endeavors (and anything else I've ever been excited about). I'm especially grateful for the constant support and encouragement of my grandma, Dolores, my parents, Steve and Leslie, and my siblings, Abbi and Stevie. I'd also like to thank my friends/chosen family, who help me find joy and optimism in the world, even during the toughest times, especially the WOA crew—Beth, Blyth, Brittawnée, and Caitlin—you mean the world to me. I'm forever grateful for all of you.

I'd be remiss not to thank my colleagues at *Environmental Health News,* whose big ideas and dedication to their work are the reasons I set out to write this book in the first place, and who have given me the rare privilege of having a job that I love. In particular, thanks to Brian Bienkowski and Douglas Fischer, who have made me a better journalist and writer, and whose support very directly made this book possible.

I'm also grateful to the professors and classmates in my MFA program at the University of San Francisco, who not only helped me hone my craft, but also gifted me the experience of being part of a community of writers—particularly Stephen Beachy, Lewis Buzbee, and Lisa Catherine Harper.

Many thanks to my editor, Emily Turner, and to the production team, the marketing team, and everyone else at Island Press who helped coax this book into the world.

Appendix

In this appendix, you'll find a list of organizations working to get cancer-causing chemicals out of various parts of our lives, loosely organized by the themes covered in each chapter. Many of these organizations work across multiple sectors and areas of interest. This list is not exhaustive, but it provides a starting point for getting involved.

Most of the organizations listed here are working at the national level. Donating time or money to the local organizations leading these efforts in your city, county, state, or region is also a highly effective way to drive change, and local groups often have fewer resources than national ones, so every contribution matters. These groups are myriad and may evolve or change names over time, so it's not practical to try to list them all here, but finding the groups doing this work in your area is a quick online search away. Some of these national groups also have regional chapters.

Food and Water
  Beyond Pesticides
  Cancer Free Economy Network
  Center for Health, Environment and Justice

Clean Water Action
Coming Clean
EarthJustice
Environmental Justice Health Alliance for Chemical Policy Reform
Environmental Protection Network
Environmental Working Group
Natural Resources Defense Council
Public Interest Research Group
Scientists, Activists, and Families for Cancer-Free Environments
    (S.A.F.E)
Sierra Club
Silent Spring Institute
Toxic-Free Future

Cosmetics and Personal Care Products

Black Women for Wellness
Breast Cancer Prevention Partners
Cancer Free Economy Network
Collaborative on Health and the Environment
Environmental Defense Fund
Environmental Working Group
Made Safe
Public Interest Research Group
Silent Spring Institute
Toxic-Free Future
WE ACT for Environmental Justice
Women's Voices for the Earth

Daycares and Schools

Child Care Aware of America
Children's Environmental Health Network

Madelina, the little girl from the introduction whose photo went viral, is among the survivors. When I first spoke with her mom, Kristin, in 2019, Madelina was a healthy, happy-go-lucky kindergarten student who loved ballet and gymnastics, riding her bike, swimming, and playing outside with her older sister. No one would ever look at her and guess that she'd been diagnosed with cancer before even reaching her second birthday.

But even when treatment is successful, it's never easy. Children often lose their hair and endure side effects like nausea, vomiting, pain, behavioral problems, and anxiety. Just as Berry feels that she'll always be haunted by breast cancer, the effects of enduring childhood cancer often last a lifetime—for both survivors and their families.

"Surviving cancer is life-changing for the patient and the family," said Dr. Erika Friehling, a pediatric hematologist/oncologist at the University of Pittsburgh Medical Center's Children's Hospital, where Madelina received her treatment. "Even seeing patients that completed treatment five to ten years ago, they still carry many of those emotions. Some parents talk about a kind of PTSD phenomenon every time they come back to Children's Hospital."

Kristin said she gets that—she and her husband find themselves holding their breath as they await the results of Madelina's regular cancer screenings, overwhelmed with gratitude each time one comes back negative.

In addition to the emotional strain, there's a significant physical one: 60% of children who survive cancer experience health problems related to the disease or their treatment, including infertility, heart failure, and secondary cancers later in life.

"We're also learning more about the neurocognitive impacts of undergoing chemotherapy as a child," Friehling said. "Specifically for the treatment of acute lymphoblastic leukemia, we're seeing some struggles in learning and developing more complex processing skills as kids get older, so we're beginning to put an emphasis on making sure those patients get extra attention as they're being treated or after being cured."

Kristin makes sure Madelina gets tested for heart problems and cognitive issues regularly since completing her treatment. "There are a gazillion unfortunate side effects," she said. "Fortunately we haven't encountered any other than some very minor developmental delays, but as she continues to grow, they're definitely something we need to watch out for."

As a result of both the growth in childhood cancer and ever-improving medical advances, one in every 530 American adults ages 20–39 has survived childhood cancer—and each faces the potential for lingering side effects for the rest of their lives. And that's just childhood cancer survivors. The American Cancer Society estimated that there were 16.9 million cancer survivors in the US as of 2019, or about 5% of the total population, and that by 2030 the total number of American cancer survivors will increase by more than 31%, to 22.1 million.

Cancer survivors are often encouraged to eat "clean" and avoid chemicals in their daily lives to help prevent their cancers from coming back. This can help some survivors feel empowered to protect themselves and may improve their health and future prognoses. But it can also make some survivors feel an unwarranted sense of guilt over the role their consumer choices may have played in their development of the disease, or feel a sense of failure if their cancer does return. And no matter how carefully survivors shop, it's just not possible to avoid carcinogenic chemicals entirely, because they're ubiquitous.

True empowerment would mean everyone being equally protected from exposure to cancer-causing chemicals. A successful war on cancer would do more than just increase the odds of survival—it would reduce the number of people ever forced to battle the disease. More than fifty years into this war, it is long past time to acknowledge that winning requires going on the offensive.

The "Personal Responsibility" Myth

The idea that our individual choices as consumers alone determine the health of the planet—and in turn, the people who inhabit it—didn't take hold by chance. Companies that extract fossil fuels and rely on plastic packaging made concerted efforts to convince consumers that dealing with cancer-causing waste is solely the public's responsibility—effectively letting industry off the hook for generating it.

In the 1950s a group of companies that were beginning to rely more heavily on plastic packaging, including Coca-Cola, PepsiCo, and Anheuser-Busch (among others), formed a nonprofit called "Keep America Beautiful." Its purported mission was to encourage environmental stewardship, but its ulterior motive was to shift responsibility for plastic waste from the companies themselves to their customers. In 1971 with help from the Ad Council, Keep America Beautiful put out its infamous "Crying Indian" TV commercial, which showed an "Indian"—actually an Italian American actor with a feather stuck in his hair—crying over the sight of trash on the ground. This supposed public service announcement was, in fact, incredibly effective propaganda: It leveraged the American public's collective guilt about our nation's history of violence against Native Americans to convince us that if the new, disposable packaging generating billions of dollars in profit for large corporations littered the environment, we had only ourselves to blame. Meanwhile, behind the scenes, the group fought legislation that would have made plastic producers more responsible for its disposal—measures that could have ameliorated the plastic pollution crisis that's now been escalating unabated for more than fifty years. Keep America Beautiful's initiatives have been described as America's first foray into corporate greenwashing, a precursor to decades of marketing spin by polluting industries.[2]

Fossil fuel companies have followed a similar blueprint, pushing personal responsibility for the climate crisis through the idea of a "carbon footprint." British Petroleum, or BP, invented the term in the early 2000s, unveiling a "carbon footprint calculator" that encouraged consumers to think about how their daily behaviors were responsible for heating the globe. Meanwhile, since the 1980s, just 100 fossil fuel producers—including BP—have been responsible for about 70% of global climate-warming emissions. BP is near the top of the list of the highest-emitting companies in the world, responsible for more than 34 billion metric tons of carbon emissions since 1965. In marketing materials, the company urged consumers to consider going on a "low-carbon diet" and touted its own forays into renewable energy, while investing less than 2.3% of its 2018 budget in renewables. Other companies jumped on the bandwagon and started pushing consumers to mind their carbon footprint—a concept that took deep root in our public discourse. Today, the term is everywhere, used by the likes of the US Environmental Protection Agency, the *New York Times*, and countless brands.[3]

These tactics seem especially egregious given that fossil fuels and their derivatives are responsible for much of our daily exposure to carcinogens. These include air pollution from the burning of fossil fuels to generate energy, vehicle emissions, petroleum-derived pesticides, dozens of petroleum-derived carcinogenic ingredients that make their way into personal care products, and chemicals that can leach into food through plastic packaging. The companies profiting from the continued proliferation of fossil fuels and plastics are not only driving climate change but also contributing to cancer cases across the globe.

When I imagine these corporate titans and PR execs, it's hard not to picture a cartoon villain cackling maliciously over his piles of money. But I know the reality is much more mundane: short-sighted decisions made by many corporate employees who are doing their jobs exception-ally well, maximizing shareholder profits with little regard for anything else. As much as corporate greed, the consequences represent profound regulatory failure—ultimately traceable to a deeply flawed political sys-tem without effective campaign finance laws. If citizens are responsi-ble for the carcinogens in our environment, it's not because we made the wrong consumer decisions, but because we—collectively—failed to properly govern.

Shifting Our Attention to Create Change

The idea that our everyday habits alone determine our cancer risk is similar to the myths about individuals bearing total responsibility for the plastic and climate crises. Our choices as individual consumers do matter when it comes to our health, but we can't solve these problems by changing those behaviors alone. We have to change our regulations to protect everyone from harmful chemicals, minimize the influence of large corporations over policymakers, and force corporations to pay for the true cost of their operations instead of continuing to externalize them at the expense of our planet and our health.

Hundreds of thousands of people around the world die from lung cancer caused by air pollution annually, and it's estimated that more than 100,000 lifetime cancer cases in the US alone could be attributed to carcinogens in tap water. And pollution in air and water account for only a fraction of the harmful exposures we experience every day. How different would our world look if we could put names and faces to the hundreds of thousands of lives that would be saved if just those two sources of exposure were eliminated through protective policies and regulations?

There are lots of ways to effectively push for meaningful, systemic change. Here are a few:

- Donate time or money to organizations that are already engaged in this fight (see the appendix for a list).
- Follow these organizations on social media, and answer and amplify their calls to action in the form of petitions and outreach to lawmakers about pending legislation.
- Sign up for notifications about proposed federal and state regulations related to carcinogens in the environment so you can contact decision makers about them when they're up for consideration (some websites where you can do this: govtrack.us, congress.gov, openstates.org, ncsl.org).
- Regularly remind your elected officials that this issue is important to their constituents.
- Amplify important stories about getting carcinogens out of the environment by sharing them and talking about them.
- Create market pressure by communicating to corporations that this issue is important to consumers. For example, you might tell your favorite microwave popcorn manufacturer that you have stopped buying their product because of concerns about PFAS in their popcorn bags, or tell a makeup brand that you've switched to their products because you appreciate their commitment to using safe ingredients.
- Support efforts aimed at campaign finance reform, restricting the power of lobbyists, and increasing transparency in the US government. (Organizations leading this work are also included in the appendix.)

If we all take even a sliver of the energy we devote to lowering our individual carbon footprint, being diligent recyclers, and reducing our personal cancer risk, and channel it toward implementing systemic changes that will protect everyone from cancer, we'll get a lot further a lot faster.

"In the end, the thing that usually gets harmful chemicals out of the public domain is public outcry," Kripke said. "Public awareness is really increasing about the fact that there are bad things in the environment we could get rid of that might make a difference right now. Knowing that gives me hope for the future."

Acknowledgments

I'd like to thank the scientists, researchers, activists, doctors, parents, and advocates who strive every day to make the world a safer, healthier place, and who generously shared their insights and ideas for this book.

Thanks also to my early readers for this book, Hattie Fletcher, Polly Hoppin, Nick Krieger, Jon Meck, and Pete Sheehy, whose feedback was invaluable. Thank you for the time, care, thought, and attention you devoted to my work, and the thoughtfulness with which you shared suggestions. My early interviews with Polly Hoppin about the Cancer Free Economy Network for *Environmental Health News* stories sparked many of the epiphanies that made me want to write about cancer and the environment, and despite being a powerhouse with a perpetually full plate, Polly has always been generous with her time, for which I am profoundly grateful.

I owe many thanks to my husband, Michael, whose love, support, and expert cooking skills have nourished me enough to make this project possible, and to our dog, Mochi, whose cuddles and shenanigans make every day better. I also owe many thanks to my family, who have always encouraged me in my writing endeavors (and anything else I've ever been excited about). I'm especially grateful for the constant support and encouragement of my grandma, Dolores, my parents, Steve and Leslie, and my siblings, Abbi and Stevie. I'd also like to thank my friends/chosen family, who help me find joy and optimism in the world, even during the toughest times, especially the WOA crew—Beth, Blyth, Brittawnée, and Caitlin—you mean the world to me. I'm forever grateful for all of you.

I'd be remiss not to thank my colleagues at *Environmental Health News,* whose big ideas and dedication to their work are the reasons I set out to write this book in the first place, and who have given me the rare privilege of having a job that I love. In particular, thanks to Brian Bienkowski and Douglas Fischer, who have made me a better journalist and writer, and whose support very directly made this book possible.

I'm also grateful to the professors and classmates in my MFA program at the University of San Francisco, who not only helped me hone my craft, but also gifted me the experience of being part of a community of writers—particularly Stephen Beachy, Lewis Buzbee, and Lisa Catherine Harper.

Many thanks to my editor, Emily Turner, and to the production team, the marketing team, and everyone else at Island Press who helped coax this book into the world.

Appendix

In this appendix, you'll find a list of organizations working to get cancer-causing chemicals out of various parts of our lives, loosely organized by the themes covered in each chapter. Many of these organizations work across multiple sectors and areas of interest. This list is not exhaustive, but it provides a starting point for getting involved.

Most of the organizations listed here are working at the national level. Donating time or money to the local organizations leading these efforts in your city, county, state, or region is also a highly effective way to drive change, and local groups often have fewer resources than national ones, so every contribution matters. These groups are myriad and may evolve or change names over time, so it's not practical to try to list them all here, but finding the groups doing this work in your area is a quick online search away. Some of these national groups also have regional chapters.

Food and Water
  Beyond Pesticides
  Cancer Free Economy Network
  Center for Health, Environment and Justice

Clean Water Action
Coming Clean
EarthJustice
Environmental Justice Health Alliance for Chemical Policy Reform
Environmental Protection Network
Environmental Working Group
Natural Resources Defense Council
Public Interest Research Group
Scientists, Activists, and Families for Cancer-Free Environments
    (S.A.F.E)
Sierra Club
Silent Spring Institute
Toxic-Free Future

Cosmetics and Personal Care Products

Black Women for Wellness
Breast Cancer Prevention Partners
Cancer Free Economy Network
Collaborative on Health and the Environment
Environmental Defense Fund
Environmental Working Group
Made Safe
Public Interest Research Group
Silent Spring Institute
Toxic-Free Future
WE ACT for Environmental Justice
Women's Voices for the Earth

Daycares and Schools

Child Care Aware of America
Children's Environmental Health Network

Environment & Human Health, Inc.
Environmental Law Institute
Healthy Schools Network
Helen R. Walton Children's Enrichment Center
National Association for Family Child Care
National Association for Regulatory Administration

Building Materials
Center for Maximum Potential Building Materials
Clean Production Action
Enterprise Green Community Guidelines
Green Science Policy Institute
Health Product Declaration Collaborative
Healthy Affordable Materials Project
Healthy Building Network
Healthy Materials Lab at Parsons School of Design
International WELL Building Institute
Living Building Challenge
Living Product Challenge
Mindful Materials
National Center for Healthy Housing
US Green Building Council

Medical Care
Alliance of Nurses for a Healthy Environment
Healthcare Without Harm
National Medical Association
Practice Greenhealth

Communities/Environmental Justice
Center for Diversity and the Environment

Center for Environmental Health
Center for Health, Environment and Justice
Clean Water Action
Climate Justice Alliance
Coming Clean
Community Action Works
EarthJustice
Environmental Justice for All
Environmental Justice Health Alliance
Global Greengrants Fund
Greenaction for Health and Environmental Justice
Honor the Earth
Indigenous Environment Network
Intersectional Environmentalist
Physicians for Social Responsibility
NAACP Environmental and Climate Justice Program
National Black Environmental Justice Network
National Environmental Law Center
Natural Resources Defense Council
The Solutions Project
WE ACT for Environmental Justice

Campaign Finance Reform/Improved Democracy

Alliance for Justice
American Civil Liberties Union
Brennan Center for Justice
Campaign Disclosure Project
Campaign Legal Center
Common Cause
Democracy 21
Free Speech For People

Issue One
Move to Amend
Open Secrets
Public Citizen
RepresentUs
Sunlight Foundation

Notes

Introduction

1. LaSalle D. Leffall and Margaret L. Kripke, *Reducing Environmental Cancer Risk: What We Can Do Now* (Washington, DC: The President's Cancer Panel, 2008), https://deainfo.nci.nih.gov/advisory/pcp/annualReports /pcp08-09rpt/PCP_Report_08-09_508.pdf.

2. Todd P. Whitehead et al., "Childhood Leukemia and Primary Prevention," *Current Problems in Pediatric and Adolescent Health Care* 46, no. 10 (2016.): 317–52, https://doi.org/10.1016/j.cppeds.2016.08.004.

3. "Key Statistics for Childhood Cancers," American Cancer Society, accessed January 12, 2021, https://www.cancer.org/cancer/cancer-in -children/key-statistics.html.

4. Madeline Drexler, "The Cancer Miracle Isn't a Cure: It's Prevention," *Harvard Public Health Magazine*, October 8, 2019, https://www.hsph .harvard.edu/magazine/magazine_article/the-cancer-miracle-isnt-a-cure -its-prevention/.

5. William H. Goodson et al., "Testing the Low Dose Mixtures Hypothesis from the Halifax Project," *Reviews on Environmental Health* 35, no. 4 (August 24, 2020): 333–57, https://doi.org/10.1515/reveh-2020-0033.

6. Molly Jacobs, Polly Hoppin, and Maggie Kuzemchak, *Environmental Chemicals and Cancer, a Science Companion Document* (Pittsburgh, PA: Cancer and Environment Network of Southwestern Pennsylvania, 2021),

https://censwpa.org/wp-content/uploads/2021/07/Science-Companion
-Document.pdf.

7. Bethsaida Cardona and Ruthann A. Rudel, "Application of an In Vitro Assay to Identify Chemicals That Increase Estradiol and Progesterone Synthesis and Are Potential Breast Cancer Risk Factors," *Environmental Health Perspectives* 129, no. 7 (July 21, 2021): 077003, https://doi.org /10.1289/ehp8608.

8. Tim Lobstein and Kelly D. Brownell, "Endocrine-Disrupting Chemicals and Obesity Risk: A Review of Recommendations for Obesity Prevention Policies," *Obesity Reviews* 22, no. 11 (August 18, 2021), https://doi.org/10 .1111/obr.13332.

9. Piera M. Cirillo et al., "Grandmaternal Perinatal Serum DDT in Relation to Granddaughter Early Menarche and Adult Obesity: Three Generations in the Child Health and Development Studies Cohort," *Cancer Epidemiology Biomarkers & Prevention* 39, no. 8 (August 1, 2021), https://doi.org /10.1158/1055-9965.epi-20-1456.

10. "Impact of EDCs on Hormone-Sensitive Cancer," Endocrine Society, accessed January 14, 2021, https://www.endocrine.org/topics/edc/what -edcs-are/common-edcs/cancer.

11. Christopher Tessum et al., "PM2.5 Polluters Disproportionately and Systemically Affect People of Color in the United States," *Science Advances* 7, no. 18 (April 28, 2021): eabf4491, https://doi.org/10.1126/sciadv .abf4491.

12. Douglas Hanahan and Robert A. Weinberg, "Hallmarks of Cancer: The Next Generation," *Cell* 144, no. 5 (March 4, 2011): 646–74, https://doi .org/10.1016/j.cell.2011.02.013.

Chapter 1. Laurel: Safer Nourishment through Science

1. Herbert P. Susmann et al., "Dietary Habits Related to Food Packaging and Population Exposure to PFASs," *Environmental Health Perspectives* 127, no. 10 (October 9, 2019), https://doi.org/10.1289/EHP4092.

2. Nathan Donley, "The USA Lags Behind Other Agricultural Nations in Banning Harmful Pesticides," *Environmental Health* 18, no. 44 (2019), https://doi.org/10.1186/s12940-019-0488-0.

3. Sona Scsukova, Eva Rollerova, and Alzbeta Bujnakova Mlynarcikova, "Impact of Endocrine Disrupting Chemicals on Onset and Development

of Female Reproductive Disorders and Hormone-Related Cancer," *Reproductive Biology* 16, no. 4 (December 2016): 243–54, https://doi.org/10.1016/j.repbio.2016.09.001.

4. Maria De Falco and Vincenza Laforgia, "Combined Effects of Different Endocrine-Disrupting Chemicals (EDCs) on Prostate Gland," *International Journal of Environmental Research and Public Health* 18, no. 18 (September 16, 2021): 9772, https://doi.org/10.3390/ijerph18189772.

5. Nicholas Kristof, "Opinion: What Poisons Are in Your Body?" *New York Times*, February 23, 2018, https://www.nytimes.com/interactive/2018/02/23/opinion/columnists/poisons-in-our-bodies.html.

6. Alan S. Kaufman et al., "The Possible Societal Impact of the Decrease in U.S. Blood Lead Levels on Adult IQ," *Environmental Research* 132 (July 2014): 413–20.

7. Bruce P. Lanphear et al., "Low-Level Environmental Lead Exposure and Children's Intellectual Function: An International Pooled Analysis," *Environmental Health Perspectives* 113, no. 7 (July 1, 2005): 894–99, https://doi.org/10.1289/ehp.7688.

8. Sydney Evans et al., "PFAS Contamination of Drinking Water Far More Prevalent than Previously Reported," Environmental Working Group, accessed January 23, 2020, https://www.ewg.org/research/national-pfas-testing/.

9. Philippe Grandjean et al., "Severity of COVID-19 at Elevated Exposure to Perfluorinated Alkylates," ed. Jaymie Meliker, *PLOS ONE* 15, no. 12 (2020): e0244815, https://doi.org/10.1371/journal.pone.0244815.

10. Jamie Dewitt et al., "Op-Ed: PFAS chemicals—the other immune system threat," *Environmental Health News*, July 6, 2020, https://www.ehn.org/pfas-and-immune-system-2646344962.html.

11. Jay Golden et al., "Green Chemistry a Strong Driver of Innovation, Growth, and Business Opportunity," *Industrial Biotechnology* 17, no. 6 (December 13, 2021): 311–15, https://doi.org/10.1089/ind.2021.29271.jgo.

Chapter 2. Ami: Safer Beauty through Racial Justice

1. Ami R. Zota and Bhavna Shamasunder, "The Environmental Injustice of Beauty: Framing Chemical Exposures from Beauty Products as a Health Disparities Concern," *American Journal of Obstetrics and Gynecology* 217,

no. 4 (October 1, 2017): 418.e1–6, https://doi.org/10.1016/j.ajog.2017
.07.020.

2. "Big Market for Black Cosmetics, but Less-Hazardous Choices Limited,"
Environmental Working Group, accessed January 12, 2022, https://www
.ewg.org/research/big-market-black-cosmetics-less-hazardous-choices
-limited.

3. Antonia M. Calafat et al., "Urinary Concentrations of Four Parabens in
the US Population: NHANES 2005–2006," *Environmental Health Perspectives* 118, no. 5 (May 1, 2010): 679–85, https://doi.org/10.1289/ehp
.0901560.

4. Kimberly Berger et al., "Personal Care Product Use as a Predictor of Urinary Concentrations of Certain Phthalates, Parabens, and Phenols in the
HERMOSA Study," *Journal of Exposure Science & Environmental Epidemiology* 29, no. 1 (January 9, 2018): 21–32, https://doi.org/10.1038
/s41370-017-0003-z.

5. Anna Maria Wróbel and Ewa Łucja Gregoraszczuk, "Actions of Methyl-,
Propyl- and Butylparaben on Estrogen Receptor-α and -β and the Progesterone Receptor in MCF-7 Cancer Cells and Non-cancerous MCF-10A
Cells," *Toxicology Letters* 230, no. 3 (November 4, 2014): 375–81, https://
doi.org/10.1016/j.toxlet.2014.08.012.

6. Ami R. Zota and Tracey J. Woodruff, "Changing Trends in Phthalate
Exposures: Zota and Woodruff Respond," *Environmental Health Perspectives* 122, no. 10 (October 1, 2014), https://doi.org/10.1289/ehp
.1408629r.

7. Chan Jin Park et al., "Sanitary Pads and Diapers Contain Higher Phthalate Contents than Those in Common Commercial Plastic Products,"
Reproductive Toxicology 84 (March 2019): 114–21, https://doi.org/10
.1016/j.reprotox.2019.01.005.

8. Chong-Jing Gao and Kurunthachalam Kannan, "Phthalates, Bisphenols,
Parabens, and Triclocarban in Feminine Hygiene Products from the
United States and Their Implications for Human Exposure," *Environment
International* 136 (March 2020): 105465, https://doi.org/10.1016/j
.envint.2020.105465.

9. Francesca Branch et al., "Vaginal Douching and Racial/Ethnic Disparities
in Phthalates Exposures among Reproductive-Aged Women: National
Health and Nutrition Examination Survey 2001–2004," *Environmental*

Health 14, no. 1 (July 15, 2015), https://doi.org/10.1186/s12940-015 -0043-6.

10. Ami R. Zota et al., "Elevated House Dust and Serum Concentrations of PBDEs in California: Unintended Consequences of Furniture Flammability Standards?," *Environmental Science & Technology* 42, no. 21 (October 1, 2008): 8158–64, https://doi.org/10.1021/es801792z.

11. Ami R. Zota et al., "Polybrominated Diphenyl Ethers, Hydroxylated Polybrominated Diphenyl Ethers, and Measures of Thyroid Function in Second Trimester Pregnant Women in California," *Environmental Science & Technology* 45, no. 18 (2011): 7896–7905, https://doi.org/10.1021 /es200422b.

12. Scott Faber, "On Cosmetics Safety, US Trails More than 40 Nations," Environmental Working Group, March 20, 2019, https://www.ewg.org /news-and-analysis/2019/03/cosmetics-safety-us-trails-more-40-nations.

13. Nirmita Panchal et al., "The Implications of COVID-19 for Mental Health and Substance Use," The Henry J. Kaiser Family Foundation, February 10, 2021, https://www.kff.org/coronavirus-covid-19/issue-brief /the-implications-of-covid-19-for-mental-health-and-substance-use/.

14. "Benefits and Costs of the Clean Air Act 1990–2020, the Second Prospective Study," US Environmental Protection Agency, Office of Air and Radiation, July 8, 2015, https://www.epa.gov/clean-air-act-overview /benefits-and-costs-clean-air-act-1990-2020-second-prospective-study.

15. Ami R. Zota and Brianna N. VanNoy, "Integrating Intersectionality into the Exposome Paradigm: A Novel Approach to Racial Inequities in Uterine Fibroids," *American Journal of Public Health* 111, no. 1 (2021): 104–9, https://doi.org/10.2105/ajph.2020.305979.

16. Mark A. Hayden et al., "Clinical, Pathologic, Cytogenetic, and Molecular Profiling in Self-Identified Black Women with Uterine Leiomyomata," *Cancer Genetics* 222 (April 1, 2018): 1–8, https://doi.org/10.1016/j.can cergen.2018.01.001.

17. Zota and VanNoy, "Integrating Intersectionality," e3.

18. Liam Downey and Brian Hawkins, "Race, Income, and Environmental Inequality in the United States," *Sociological Perspectives* 51, no. 4 (December 1, 2008): 759–81, https://doi.org/10.1525/sop.2008.51.4.759.

19. Haley M. Lane et al., "Historical Redlining Is Associated with Present-Day Air Pollution Disparities in U.S. Cities," *Environmental Science &*

Technology Letters 9, no. 4 (March 9, 2022): 345–50, https://doi.org/10
.1021/acs.estlett.1c01012.

Chapter 3. Nse: Safer Little Ones through Politics

1. Ben Daniel Spycher et al., "Parental Occupational Exposure to Benzene and the Risk of Childhood Cancer: A Census-Based Cohort Study," *Environment International* 108 (November 2017): 84–91, https://doi.org/10 .1016/j.envint.2017.07.022.

2. Simon Chang et al., "DDT Exposure in Early Childhood and Female Breast Cancer: Evidence from an Ecological Study in Taiwan," *Environment International* 121, part 2 (December 2018): 1106–12, https://doi .org/10.1016/j.envint.2018.10.023.

3. Michael C. R. Alavanja and Matthew R. Bonner, "Occupational Pesticide Exposures and Cancer Risk: A Review," *Journal of Toxicology and Environmental Health* Part B, Critical Reviews 15, no. 4 (May 9, 2012): 238–63, https://doi.org/10.1080/10937404.2012.632358.

4. Geneviève Van Maele-Fabry, Perrine Hoet, and Dominique Lison, "Parental Occupational Exposure to Pesticides as Risk Factor for Brain Tumors in Children and Young Adults: A Systematic Review and Meta-Analysis," *Environment International* 56 (June 2013): 19–31, https://doi .org/10.1016/j.envint.2013.02.011.

5. Darryl Fears, "The president just signed a law that affects nearly every product you use," *Washington Post*, June 22, 2016, https://www.washing tonpost.com/news/speaking-of-science/wp/2016/06/22/obamatoxic/.

Chapter 4. Bill: Safer Homes and Offices through Market Pressure

1. "Vinyl Chloride—Cancer-Causing Substances," National Institutes of Health, National Cancer Institute, accessed January 30, 2022, https:// www.cancer.gov/about-cancer/causes-prevention/risk/substances/vinyl -chloride.

2. "Victims of Wood Preservatives Want Them Fully Banned/EPA Announces Phase Out of Residential Uses of CCA-Treated Wood," Beyond Pesticides, accessed November 29, 2021, https://www.beyond pesticides.org/programs/wood-preservatives/media/no-title89.

3. "Removing Arsenic from Building Materials: A Success Story," Healthy Building Network, accessed January 26, 2022, https://healthybuilding .net/blog/8-removing-arsenic-from-building-materials-a-success-story.

4. Molly Jacobs et al., *Review of Trends in the Use and Release of Carcinogens in Massachusetts* (Lowell, MA: Toxics Use Reduction Institute, 2013), https://www.turi.org/content/download/7915/138160/file/Carcinogens%20Report.pdf.

5. Bárbara Pinho, "Whatever Happened to the Hole in the Ozone Layer?" *Discover Magazine*, November 10, 2020, https://www.discovermagazine.com/environment/whatever-happened-to-the-hole-in-the-ozone-layer.

6. James Vallette, "Review of Residential Fiberglass Insulation Transformed: Formaldehyde Is No More," Healthy Building Network, October 30, 2015, https://healthybuilding.net/blog/204-residential-fiberglass-insulation-transformed-formaldehyde-is-no-more.

7. Helene Wiesinger, Zhanyun Wang, and Stefanie Hellweg, "Deep Dive into Plastic Monomers, Additives, and Processing Aids," *Environmental Science & Technology* 55, no. 13 (June 21, 2021): 9339–51, https://doi.org/10.1021/acs.est.1c00976.

Chapter 5. B. Braun: Safer Medical Treatment through Innovation

1. Tsung-Hua Hsieh et al., "DEHP Mediates Drug Resistance by Directly Targeting AhR in Human Breast Cancer," *Biomedicine & Pharmacotherapy* 145 (January 2022): 112400, https://doi.org/10.1016/j.biopha.2021.112400.

2. Matthew Genco, Lisa Anderson-Shaw, and Robert M Sargis, "Unwitting Accomplices: Endocrine Disruptors Confounding Clinical Care," *Journal of Clinical Endocrinology & Metabolism* 105, no. 10 (October 2020): e3822–27, https://doi.org/10.1210/clinem/dgaa358.

3. Kristina M. Zierold and Clara G. Sears, "Are Healthcare Providers Asking about Environmental Exposures? A Community-Based Mixed Methods Study," *Journal of Environmental and Public Health* 2015 (October 8, 2015): e189526, https://doi.org/10.1155/2015/189526.

4. Roland Geyer, "Production, Use, and Fate of Synthetic Polymers," in *Plastic Waste and Recycling: Environmental Impact, Societal Issues, Prevention, and Solutions*, ed. Trevor M. Letcher (Cambridge, MA: Academic Press, 2020), 13–32, https://doi.org/10.1016/B978-0-12-817880-5.00002-5.

5. Laura Sullivan. "How Big Oil Misled the Public into Believing Plastic Would Be Recycled." NPR. September 11, 2020. https://www.npr.org/2020/09/11/897692090/how-big-oil-misled-the-public-into-believing-plastic-would-be-recycled.

6. Chad Pradelli and Cheryl Mettendorf, "Federal Agency Fails to Notify Allentown Residents about Their Increased Cancer Risk," 6 ABC Philadelphia, February 11, 2022, https://6abc.com/bbraun-allentown-cancer-lawsuit-ethylene-oxide-eto-environmental-protection-agency/11552914/.

7. Binghui Huang, "EPA raises concern about elevated cancer risk for people living around B. Braun plant near Allentown," *The Morning Call,* July 20, 2019, https://www.mcall.com/health/mc-hea-cancer-risk-lehigh-20190718-sj2he3ten5cfxc6gwg54zxb3ne-story.html.

8. Sharon Lerner, "Trump EPA Invited Companies to Revise Pollution Records of a Potent Carcinogen," *The Intercept*, March 18, 2021, https://theintercept.com/2021/03/18/epa-pollution-cancer-ethylene-oxide/.

9. Lauren Vogel, "Fracking Tied to Cancer-Causing Chemicals," *Canadian Medical Association Journal* 189, no. 2 (January 16, 2017): E94–95, https://doi.org/10.1503/cmaj.109-5358.

10. Lisa M. McKenzie et al., "Childhood Hematologic Cancer and Residential Proximity to Oil and Gas Development," ed. Jaymie Meliker, *PLOS ONE* 12, no. 2 (February 15, 2017): e0170423, https://doi.org/10.1371/journal.pone.0170423.

11. Molly Jacobs, Polly Hoppin, and Maggie Kuzemchak, *Environmental Chemicals and Cancer: A Science Companion Document*, Cancer and Environment Network of Southwestern Pennsylvania, July 2021, https://censwpa.org/wp-content/uploads/2021/07/Science-Companion-Document.pdf.

12. Jamiles Lartey and Oliver Laughland, "Almost Every Household Has Someone That Has Died from Cancer," *The Guardian*, May 6, 2019, https://www.theguardian.com/us-news/ng-interactive/2019/may/06/cancertown-louisana-reserve-special-report.

13. "B. Braun Medical: Protecting Lives Here & Across the Country," accessed April 14, 2022, www.bbraungetthefacts.com.

14. Douglas Fischer, "Lawmakers urge action on safer medical products," *Environmental Health News*, February 22, 2021, https://www.ehn.org/toxic-neonatal-iv-bags-2650625484.html.

Chapter 6. Melanie: Safer Neighborhoods through Activism

1. Drew Michanowicz et al., *Pittsburgh Regional Environmental Threats Analysis (PRETA) Report*, University of Pittsburgh Graduate School of Public

Health, August 2013, https://www.heinz.org/UserFiles/Library/PRETA
_HAPS.pdf.

2. Molly Jacobs, Polly Hoppin, and Maggie Kuzemchak, *Environmental
Chemicals and Cancer: A Science Companion Document*, Cancer and Environment Network of Southwestern Pennsylvania, July 2021, https://
censwpa.org/wp-content/uploads/2021/07/Science-Companion-Docu
ment.pdf.

3. Kristina Marusic, "Kids in southwestern Pennsylvania are exposed to
carcinogenic coke oven emissions at shockingly higher rates than the rest
of the country," *Environmental Health News*, April 22, 2019, https://www
.ehn.org/us-steel-pittsburgh-cancer-2634765539.html.

4. Christopher W. Tessum et al., "PM2.5 Polluters Disproportionately and
Systemically Affect People of Color in the United States," *Science Advances*
7, no. 18 (April 28, 2011): eabf4491, https://doi.org/10.1126/sciadv
.abf4491.

5. Omar Hahad et al., "Ambient Air Pollution Increases the Risk of Cerebro-
vascular and Neuropsychiatric Disorders through Induction of Inflammation and Oxidative Stress," *International Journal of Molecular Sciences* 21,
no. 12 (June 17, 2020): 4306, https://doi.org/10.3390/ijms21124306.

6. Douglas J. Myers et al., "Letter to the Editor: Cancer Rates Not
Explained by Smoking: How to Investigate a Single County," *Environmental Health* 20, no. 1 (May 21, 2021), https://doi.org/10.1186/s12940
-021-00737-8.

7. Tricia L. Morphew et al., "Impact of a Large Fire and Subsequent Pollution Control Failure at a Coke Works on Acute Asthma Exacerbations in
Nearby Adult Residents," *Toxics* 9, no. 7 (June 25, 2021): 147, https://
doi.org/10.3390/toxics9070147.

8. Oliver Morrison, "A Series of Maintenance Failures Led to the 2018 Fire
at the Clairton Coke Works," *PublicSource*, November 19, 2021, https://
www.publicsource.org/clairton-coke-works-maintenance-us-steel
-pollution-control-fire/.

9. Kristina Marusic, "ER visits for asthma dropped 38% the year after one
of Pittsburgh's biggest polluters shut down," *Environmental Health News*,
May 8, 2018, https://www.ehn.org/shenango-coke-works-closed-asthma
-dropped-2566777141.html.

Epilogue. Moving Beyond Survival

1. "Risk of Dying from Cancer Continues to Drop at an Accelerated Pace," American Cancer Society, January 12, 2022, https://www.cancer.org /latest-news/facts-and-figures-2022.html.

2. Sharon Lerner, "How the Plastics Industry Is Fighting to Keep Polluting the World," *The Intercept*, July 20, 2019, https://theintercept.com/2019 /07/20/plastics-industry-plastic-recycling/.

3. Mark Kaufman, "The Devious Fossil Fuel Propaganda We All Use," *Mashable*, July 13, 2020, https://mashable.com/feature/carbon-foot print-pr-campaign-sham.

Index

About the Author

Kristina Marusic is an award-winning journalist who covers environmental health and justice for *Environmental Health News*. She holds an MFA in nonfiction writing from the University of San Francisco, and her personal essays and reporting on topics ranging from the environment, LGBTQ+ equality, and politics to feminism, food, and travel have been published by outlets including CNN, *Slate*, *Vice*, *Women's Health*, the *Washington Post*, MTV News, *The Advocate*, and *Bustle*, among others. She lives in Pittsburgh with her partner of ten years, Michael, and the cutest dog in the world, Mochi. You can visit her online at KristinaMarusic.com.